AF712431

Springer Series in Neuropsychology

Harry A. Whitaker, Series Editor

Springer Series in Neuropsychology

Harry A. Whitaker, Series Editor

Phonological Processes and Brain Mechanisms
H.A. Whitaker (Ed.)

Contemporary Reviews in Neuropsychology
H.A. Whitaker (Ed.)

Neuropsychological Studies of Nonfocal Brain Damage: Dementia and Trauma
H.A. Whitaker (Ed.)

Harry A. Whitaker
Editor

Contemporary Reviews in Neuropsychology

Springer-Verlag
New York Berlin Heidelberg
London Paris Tokyo

Harry A. Whitaker
Professor of Neurology
Department of Neuroscience
University of North Dakota School of Medicine
Fargo, North Dakota 58102, USA
and
Director, Cognitive Neuroscience Research Program
The Neuropsychiatric Institute
Fargo, North Dakota 58103, USA

Library of Congress Cataloging-in-Publication Data
Contemporary reviews in neuropsychology.
(Springer series in neuropsychology)
Includes bibliographies and index.
1. Neuropsychology. 2. Brain damage.
3. Aphasia. I. Whitaker, Harry A. II. Series.
[DNLM: 1. Neuropsychology. WL 103 C761]
QP360.C663 1988 616.8 87-26394

© 1988 by Springer-Verlag New York Inc.
All rights reserved. This work may not be translated or copied in whole or in part without the written permission of the publisher (Springer-Verlag, 175 Fifth Avenue, New York, New York 10010, USA), except for brief excerpts in connection with reviews or scholarly analysis. Use in connection with any form of information storage and retrieval, electronic adaptation, computer software, or by similar or dissimilar methodology now known or hereafter developed is forbidden.
The use of general descriptive names, trade names, trademarks, etc. in this publication, even if the former are not especially identified, is not to be taken as a sign that such names, as understood by the Trade Marks and Merchandise Marks Act, may accordingly be used freely by anyone.
While the advice and information in this book are believed to be true and accurate at the date of going to press, neither the authors nor the editors nor the publisher can accept any legal responsibility for any errors or omissions that may be made. The publisher makes no warranty, express or implied, with respect to the material contained herein.

Typeset by Best-set Typesetter Ltd., Chai Wan, Hong Kong.
Printed and bound by Edwards Brothers, Inc., Ann Arbor, Michigan.
Printed in the United States of America.

9 8 7 6 5 4 3 2 1

ISBN 0-387-96606-4 Springer-Verlag New York Berlin Heidelberg
ISBN 3-540-96606-4 Springer-Verlag Berlin Heidelberg New York

Preface: The Scope of Experimental Neuropsychology

Traditional experimental neuropsychology had been concerned with focal brain lesions, the functions of discrete parts of brain, and modular components of language and memory. Over the last two decades, an evolving cognizance of right hemisphere functional capacities has led to a reexamination of simple models of laterality and an interest in other behaviors besides language and memory. Concomitant with the expansion of topics has been an increased experimental and theoretical sophistication, no doubt partly due to the intermingling of disciplines from medicine, social and behavioral science, information and computer science, and the arts and humanities.

From a practical standpoint, the maturation of neuropsychology has also been marked by the introduction of new journals, the expansion of existing ones, and a plethora of new books. At least one handbook, in two volumes, is available, as is a handbook of assessment instruments. The major professional society in neuropsychology is undertaking the production of an encyclopedia of the field. The only publication-type apparently missing from this development is an appropriate forum for the scholarly review—the review that exceeds the customary journal article in length and depth, but falls short of the monograph. The field does not yet have a publication that approximates an annual review. The present volume is a test of the viability of the review article format; each chapter does not closely tether to a common theme, although each does capture the essence of the scope of contemporary neuropsychology. The first two chapters consider issues in laterality, the first proposing a model for lateral asymmetries and the second proposing a model for left and right hemisphere contributions to lexical processing. The third chapter is a comparative study of the neural mechanisms of spatial processing in man and infrahuman species. The fourth chapter addresses the question of humor as a right hemisphere-mediated function, at the interface between affective behavior and high level language processing. The last chapter addresses an old and unresolved modularity problem in neuropsychology, the question of whether aphasic patients also sustain an impairment in intelligence. A precis of each chapter follows.

Bradshaw, Pierson-Savage, and Nettleton present a hemisphere-hemispace mapping account of asymmetries as a refinement of the anatomic connectivity model. They begin with a review of hemi- or unilateral neglect, noting the five components to this syndrome: hemi-inattention, hemi-spatial neglect, hemi-akinesia, allesthesia, and extinction. Neglect should be considered in reference to several space coordinate systems: body midline, head coordinates, and retinal coordinates. These three systems may be dissociated by appropriate turning manipulations. A review of research on lesion localization and lateralization and hemispheric mechanisms suggests that right hemisphere, post-Rolandic lesions are the most frequent cause of neglect. Bradshaw, Pierson-Savage, and Nettleton then turn to normal subjects, experimental studies of line bisection, and visual field (tachistoscopic) and ear advantage (dichotic) asymmetries that are affected by manipulations of hemispace; these include kinesthetic studies of magnitude estimation in which hand and hemispace are unconfounded, visual studies in which hemi-retinal and hemi-spatial variables are dissociated, and vibrotactile studies that address the effect of hemispace as opposed to anatomic connectivity. They conclude that the position in extracorporeal space in which events occur, or seem to occur, is as important in determining lateral asymmetries as are the basic anatomic connections—ear of entry, visual field or hand. Hemispace asymmetries are manifested both as underestimation of magnitudes to the left side and as superior performance in processing to the left or right side. An interesting set of studies of auditory hemispace created an analog to the dichotic right ear advantage (REA) in terms of the *perceived* spatial location of a sound source, not its actual location, thus showing that part of the REA is not due to anatomic connections, but to a right hemispace superiority for auditory processing. Extending these studies to brain-damaged subjects, Bradshaw, Pierson-Savage, and Nettleton were able to separate the left hemi-neglect due to right hemisphere lesions from visually based compensatory strategies that patients had acquired during rehabilitation. Hemispace asymmetries seem to be stronger than anatomic pathway effects, which present some interesting consequences for theories of handedness. Based on the evidence that the perceived spatial location of a stimulus and the perception of the individual's position in space are the product of integrating complex multimodal stimuli, Bradshaw, Pierson-Savage, and Nettleton suggest that the brain maintains two internal maps. One represents the proximal receptor surfaces and is responsible for conventional lateral asymmetries. The other represents extracorporeal space and is responsible for hemispace asymmetries. They conclude by suggesting that the traditional anatomic connectivity model cannot explain the kind of data that show that spatial location may determine performance asymmetries. Furthermore, the fact that visual field and ear asymmetries readily occur to (functionally) appropriate lateralized stimuli, which are randomly presented to the left or right side, shows that asymmetries cannot solely result from attentional factors. The best model seems to be one that posits a functional relationship between each hemisphere and its contralateral hemispace, and includes a role for attentional factors as well.

Chiarello reviews the complex and voluminous literature on visual half-field studies of lexical processing. These studies use three general tasks: naming, lexical decision and word judgments. First, it appears that the left hemisphere (LH) and right hemisphere (RH) are equally proficient at visual-sensory feature detection; there is little evidence that LH-RH asymmetries in lexical processing may be attributed to this early stage of processing. When words are displayed in the normal horizontal manner in normal typeface, there is a length effect in the left visual field (LVF), indicating that the RH processes words serially, letter-by-letter. At the lexical access stage, word frequency effects are equivalent in the LH and RH. The data supporting a hemisphere asymmetry in lexical access for concrete versus abstract words are quite equivocal and, in fact, there are no convincing data to suggest that certain classes of words are any less accessible than others when presented to the RH. In other words, there is little evidence that unequal lexicons are available to the normal LH and RH. However, whether the lexicon is actually *located in* the LH or the RH remains to be demonstrated; visual half-field studies do not resolve this issue. Some postlexical access processing studies have pointed to a LH advantage (a) when words are matched on the basis of semantic category membership, (b) when processing depends on the retrieved meaning of abstract words, and (c) when processing requires phonetic comparisons of lexically retrieved information. Chiarello concludes that the left and right hemispheres have different reading profiles—the LH encodes words by a direct specialized process (whole word), while the RH encodes words serially using a letter-by-letter process. Despite the prelexical differences, both hemispheres appear to be equally able to access semantic information; thus, current evidence does not support the idea that the RH accesses an abridged, concrete, imageable lexicon, biased to particular word classes. On the other hand, postlexical access asymmetries do exist, favoring the LH. Contrary to Coltheart's model, there is no evidence that the RH reading profile resembles deep dyslexia or, for that matter, any acquired dyslexic syndrome.

Beatty and Tröster critically review the proposed neural mechanisms underlying spatial behavior in man and infrahuman species. The data are presented in relation to different brain regions, beginning with an analysis of the contribution of the right temporal lobe to spatial learning and memory. Two models of hippocampal function are reviewed: the working memory and the cognitive map hypotheses. The contribution of the frontal lobe to spatial behavior is reviewed with respect to the notions of egocentric versus allocentric orientation, rather than spatial memory. As is well known, posterior neocortex lesions give rise to a great variety of visuoperceptual and visuospatial disorders, making them quite difficult to interpret in terms of learning and memory. Beatty and Tröster next review the role in spatial memory of two subcortical areas; the caudate and the medial diencephalon. Caudate lesions in general result in deficits similar to frontal lesions: impairing performance on delayed alternation and delayed response tasks. The medial diencephalon's role in memory had been assumed because of the common association of pathology in this area with the amnesia in alcoholic Korsakoff's syndrome. However, the considerable frontal lobe damage and probable hippocampal damage in alcoholic Korsakoff (AK) patients suggest that AK amnesia should not be attributed to the medial diencephalon at all. Neuropharmacology is next reviewed by Beatty and Tröster. Central cholinergic circuits play an important role in spatial memory, e.g., anticholinergic drugs act on encoding processes, probably by blocking cholinergic input to hippocampus and frontal lobe. The last discussion in the review is on aging effects; age-related changes in spatial memory correlate with and thus may be associated with hippocampal pathology. Overall, Beatty and Tröster question the usefulness of the egocentric/allocentric distinction that is frequently utilized in spatial learning and memory research. When looked at carefully, it seems that brain damage typically effects both ego- and allo-functions. Additionally, the evidence suggests that the temporal lobe is as much involved in nonspatial as in spatial memories. Beatty and Tröster conclude that a number of cortical and subcortical regions are involved in spatial memory, although the precise roles of these regions have not yet been elucidated.

As Bihrle, Brownell, and Gardner point out, an interest in the sense of humor has both scientific and clinical parameters. In the first place, one's understanding of this important human attribute is enhanced by studying its decomposition. Second, a patient's rehabilitation can be materially influenced by the retention or loss of the sense of humor. Right hemisphere lesions have been shown to impair the sense of humor even though language function is intact. Bihrle, Brownell, and Gardner accept the incongruity-resolution theory of humor and show that RH lesioned patients are sensitive to the form of a joke—the presence of incongruity or surprise—but fail to appreciate how the incongruous elements can be integrated into a context with other elements. Nonetheless, RH-lesioned patients understand the formal differences between a straight story and a joke. Four factors that may contribute to RH-lesioned patients' problems with humor are reviewed: (a) rigidity of interpretation, (b) literalness, (c) inattention to relevant detail, and (d) impairment of affect. They conclude that the deficit in humor comprehension is a narrative one (the first three of the four factors just cited) and not an emotional one (the fourth factor) in spite of the well known evidence that RH lesions may impair affective processing.

Shortly after Broca's initial publications on aphasia, Laborde and

Trousseau discussed the impairment of intelligence in aphasia. As noted in Gainotti's review, the question is still unresolved. What is the relation between language and thought? Gainotti reviews the early history of this issue, the current research, and the various theoretical interpretations that have been offered to explain the relationship. Finkelnburg characterized aphasia as an aspect of asymbolia, a view accepted by Kussmaul and Hughlings Jackson and later developed into different versions by Marie, Head, and Goldstein. It is immediately apparent that this theme was not part of the Wernicke-Lichtheim tradition that has influenced so much of modern neuropsychology. Gainotti points out that some of the early concepts of intelligence were akin to a current view of semantic memory, e.g., our permanent knowledge about basic facts of the world. Recent research supports an intermediate view that there is a specific cognitive deficit in aphasia not attributable to either diffuse intellectual impairment nor to a linguistic impairment. De Renzi and colleagues have demonstrated that aphasic patients are consistently impaired on both intelligence (abstract problem-solving) tests and associative (concrete feature-matching) tests. However, some of the impairment is a general consequence of brain damage independent of the presence of aphasia, e.g., RH-damaged non-aphasic patients are also impaired on intelligence tests. Some of the impairment does seem specific to aphasia, e.g., the poorer performance on associative tests of feature matching. However, the correlation between severity of aphasia and severity of impaired nonverbal intelligence is generally quite variable. Nonverbal IQ is usually affected in aphasia, but does not have to be. Thus, a modular view of the LH is maintained by claiming that the language areas and the nonverbal IQ areas are anatomically contiguous. The modular model has been reinforced by recent studies on pantomime and gesture production in aphasia; again, no clear statistical relationship between gestural impairment and aphasia was observed. However, Gainotti's own research on the *comprehension* of pantomime and gesture has consistently shown a clear correlation between severity of aphasia and severity of impaired gesture and pantomime recognition and, further, this correlation is not observed in other groups of brain-damaged subjects. Unlike the previously cited research, this work supports the view that aphasia is part of a more general symbolic deficit. Some current theorists maintain a weak form of the asymbolia theory (i.e., that anatomically contiguous but functionally independent structures subsume language and nonverbal cognitive functions). Gainotti's work, on the other hand, suggests that the components of language that are affected in aphasia are linked with nonverbal cognitive systems; these components are *central* and hence are linguistic in nature. The primary candidate for this central component is the semantic-lexical level of language. Gainotti argues that this level, and not the more form-oriented levels of phonology or syntax, is the level associated with nonverbal cognition. Although this view is not in agreement with a modular view of brain organization, modularity may be saved if one maintains that nonverbal cognitive tests are covertly verbally mediated. This explains why the aphasic impairment affects nonverbal test performance. Gainotti concludes that a functional relationship between verbal and nonverbal cognitive disturbances in aphasic patients does exist. The nature of this relationship remains to be fully explicated.

The scope of experimental neuropsychology today exceeds the capacity of most researchers, a familiar problem in any mature discipline. The natural result is specialization. And the corollary of specialization is the exigency of the review article, whence the present volume.

HARRY A. WHITAKER

Contents

Contributors

William W. Beatty Department of Psychology, North Dakota State University, Fargo, North Dakota 58105, USA

Amy M. Bihrle Department of Psychology, University of California–San Diego, San Diego, California 92110, USA

John L. Bradshaw Department of Psychology, Monash University, Clayton, Victoria, Australia 3168

Hiram H. Brownell Psychology Service (116 B), V.A. Medical Center, Boston, Massachusetts 02130, USA

Christine Chiarello Department of Psychology, Syracuse University, Syracuse, New York 13244-2340, USA

Guido Gainotti Institute of Neurology, Catholic University, Policlinico A. Gemelli, Largo A. Gemelli n. 8, 00168 Rome, Italy

Howard Gardner Psychology Service (116 B), V.A. Medical Center, Boston, Massachusetts 02130, USA

Norman C. Nettleton Department of Psychology, Monash University, Clayton, Victoria, Australia 3168

Jane M. Pierson-Savage Department of Psychology, Monash University, Clayton, Victoria, Australia 3168

Alexander I. Tröster University of California–San Diego, San Diego State University Clinical Psychology Department, San Diego, California 92182-0551, USA

1
Hemispace Asymmetries

John L. Bradshaw, Jane M. Pierson-Savage, and
Norman C. Nettleton

Hemineglect: Description of a Syndrome

Unilateral brain damage may lead to (usually left) hemineglect (De Renzi, 1982; Heilman, Bowers, & Watson, 1984; Heilman & Valenstein, 1979; Mesulam, 1981, 1983). In severe cases it is as if the left half of the patient's world has ceased to exist, with failure to report, respond to, or orient towards stimuli on the side contralateral to the lesion, in the absence of primary sensory or motor defects. The patient may shave, groom or dress only the right side, leave food untouched on the left side of the plate, read only words, letters or sentences written on the right side of a page, leave an unusually wide margin on the left, copy only the right side of a drawing, and bisect a visually presented horizontal line to the right of the true midpoint, especially if the whole line lies to the left of the midline (Heilman & Valenstein, 1979). There may even be neglect of the left side of a scene that is currently being imagined (Bisiach & Luzzatti, 1978), such that the patient's inability to recall objects depends upon his or her imaginary standpoint. Thus the affected "space" with respect to the actual observer (patient) can be delineated, i.e., extrapersonal space, or with respect to the vantage point of an imagined observer during topographical recall, or with respect to a particular component of a larger object (Halsband, Gruhn, & Ettlinger, 1985). Indeed sometimes the left side of all objects in view may be neglected, and on other occasions all objects falling in the left half of the field may be ignored, indicating the importance of attentional mechanisms in this syndrome. In fact the condition is a hybrid between the lateralized phenomena of language and spatial processing on the one hand, and on the other, the largely contralateral (though also ipsilateral, in certain circumstances, see below) representation of sensory and motor events.

The lesions causing the syndrome are usually large and either of sudden onset or are rapidly progressive; Heilman, Watson, Valenstein, and Damasio (1983) suggest that they are responsible for five separable components of the syndrome:

Hemi-inattention, or failure to respond to unilateral stimuli unless attention is drawn to them; *Hemispatial neglect*, the classical defects described above; *Hemiakinesia*, failure in the absence of gross motor defects to raise an arm contralateral to the lesion, or to orient head and eyes to a stimulus contralateral to the lesion unless encouraged to do so; *Allesthesia*, where a patient touched on the left side may report its occurrence on the right side; *Extinction*, which when all the other above symptoms have finally abated, as they may do in a matter of months, may remain relatively intractable.

It is still not known whether the mechanisms of apparent recovery involve the undamaged hemisphere, ipsilateral pathways, the commissures, or subcortical areas (e.g. the superior colliculus). With extinction, a patient capable of adequate response to unilaterally presented stimuli on either side may fail to report stimuli contralateral to the damaged hemisphere when given simultaneous bilateral stimulation (visual, auditory, or tactual). With normal subjects, the dichotic right-ear advantage with competing simultaneous stimulation of both ears may be an analogous phenomenon (see Bradshaw, Burden, & Nettleton, 1986, for a review). Of course hemineglect does not depend upon extinction, and unlike right hemineglect right side extinction is by no means rare (Schwartz, Marchok, Kreinick, & Flynn, 1979). Moreover, unilateral neglect may be considered with reference to more than one coordinate system, for example the body midline (the currently preferred viewpoint), head coordinates (when the head is turned with respect to the body), and even *retinal coordinates*, with respect to gaze direction (Bisiach, Capitani, & Porta, 1985; Heilman et al., 1984); all three coordinate systems can be mutually dissociated by appropriate turn conditions. Bisiach et al. (1985; cf. Bradshaw & Pierson, 1985) have concluded that egocentric and extracorporeal space is organized topographically in the brain, in a projection system separate from that of the proximal receptor or body surface, and that there are circumscribed brain areas wherein lesions can result in representational loss limited to definite regions of space. Bisiach et al. found that the boundary of the neglected area of the tactile apparatus was influenced both by the sagittal midplane of the trunk, and by line of sight in terms of both head and gaze orientation. They also cite physiological and anatomic evidence of thalamic and premotor cells firing to visual stimuli in a definite region of peripersonal space irrespective of gaze direction, and independently of the retinal coordinates of the proximal stimulus.

Areas Affected in Hemineglect

Hemineglect is not a unitary phenomenon; different sensory modalities and different processes are disturbed in association with lesions in different cerebral loci. Important factors are locus of lesion, speed and nature of the destructive process, elapsed time between damage and testing, and premorbid processing strategies. Cortical polymodal sensory convergence areas interconnecting with the inferior parietal lobule may be involved, together with the cingulate gyrus, frontal regions, the basal ganglia, thalamus, and mesencephalic reticular formation, all of which are richly interconnected and are involved with arousal and attention to meaningful stimuli (Bisiach, Comacchia, Sterzi, & Vallar, 1984; Bisiach et al., 1985; Crowne, 1983; Heilman, Watson, & Valenstein, 1985; Luh, Butter, & Buchtel, 1986; Mesulam, 1981, 1983, 1985; Mishkin, Ungerleider, & Macko, 1983; Stuss & Benson, 1984). Pandya and Yeterian (1984) propose that visual information reaches the cortex via both tectal and thalamic routes, thus permitting both initial orientation to and identification of components in the visual world to occur. They suggest that visual and somatosensory information, after being received in the primary projection areas and being elaborated and identified in the appropriate parasensory areas, is transmitted to three post-Rolandic centers involved in spatial location. These are the intraparietal sulcus (including area POa) which is involved in the integrative scanning of the visual–somatosensory sphere; the superior temporal sulcus (including areas PGa and TPO), which integrates complex visual–somatosensory images; and the inferior parietal lobule (area 7).

Area 7 is also responsible for spatial location memory, and projects to the cingulate gyrus in the limbic regions for evaluation of sensory significance. Damage to either the inferior parietal lobule or the cingulate gyrus can result in hemineglect and a consequent loss of comprehension of the emotional significance of events in the contralateral space; a damaged inferior parietal lobule has perhaps more of an effect on outwardly oriented aspects, and a damaged cingulate gyrus on inwardly oriented motivational components. Both regions project to the prefrontal cortex and provide it with an overall sense of spatiality within which behavioral sequencing may be initiated, appropriate to the subject's current motivational state.

Frontal regions, including the arcuate gyrus (the frontal eyefields) mediate a motor map for the distribution of scanning, orienting, and exploration in extrapersonal space, with neural responses very like those of the superior colliculus. Both regions are involved with oculomotor control, though neurons in the frontal regions may respond only when eye movements occur towards important stimuli, while posterior neurons in the inferior parietal lobule may respond to important stimuli whether or not eye movements occur. Although we can therefore distinguish between the posterior polymodal cortex of the inferior parietal lobule with its sensory template of the external world and anterior regions containing a motor map for orienting within extrapersonal space, according to Ogden (1987) injury to the posterior right hemisphere (RH) is more likely to lead to severe left neglect, while injury to the anterior left hemisphere (LH) may also bring about a similar right neglect though this condition is less severe and resolves more quickly.

Finally, the cingulate gyrus, with its motivational map for the distribution of interest and expectancy, outputs to the hippocampus for the storage and retrieval of spatial memories.

Possible Mechanisms of Hemineglect

Left-sided neglect after injury to the RH is more common, severe, and longer lasting than right hemineglect after injury to the LH (Mesulam, 1981; Ratcliff, 1982; cf. Ogden, 1985); however, the latter condition may indeed be masked by incapacitating aphasia after trauma to the LH. The phenomenon of left-sided neglect cannot, however, be simply due to right hemisphere specialization for visuospatial processing because if these mechanisms were damaged we would not expect the disability to be largely limited to left hemispace. Does hemineglect result from the prevalence of the contraversial turning tendency of the intact hemisphere that can no longer be countered by the damaged hemisphere? Is neglect worse after damage to the RH because verbal exchanges between physician and patient activate the verbally controlling LH and produce an overflow of excitation from the language areas to the oculomotor centers for rightwards turning (Kinsbourne, 1970)? If so, then why does left hemineglect still occur even with such nonverbal tasks as figure copying? Is it then perhaps a matter of hemispheric asymmetry in controlling eye movements? Smith and Latto (1982) suggest that lesions in either hemisphere can result in hemineglect, but while patients with damage to the left parietal area can compensate via scanning, those with right parietal injuries cannot because of disordered eye movements. De Renzi, Colombo, Faglioni, and Gilbertoni (1982) claim that conjugate gaze paresis is more frequent, severe, and long lasting after damage to the RH, when post-Rolandic lesions are usually found; in those with damage to the LH and gaze paresis the entire territory of distribution of the Sylvian artery is usually involved. Thus they say that oculomotor centers are diffusely organized in the LH and focally in the RH, and reject the idea that conjugate gaze paresis and turning towards the side of the lesion is secondary to hemi-inattention, because such vergence may be marked in coma. However the occurrence of hemineglect in nonvisual tasks indicates that deviated eyemovements may stem from but cannot cause neglect.

A traditional view (see De Renzi et al., 1982; Geschwind, 1981; Heilman, 1979; Mesulam, 1981) is that while the LH exclusively mediates attention for contralateral (right) hemispace, the intact RH can cope with both sides, even though its dominant tendency might be towards contralateral (left) hemispace. Thus damage to the LH is unlikely to produce hemineglect, as the intact RH can take over, while damage to the RH will lead to left hemineglect as the left hemisphere cannot compensate. Certainly injury to the RH is often associated with lowered affect, defective alerting, poorer arousal and reaction time decrements (with either hand), all indicating (Heilman, 1979) that the RH may mediate bilateral as well as contralateral arousal. Damage to the RH may also lead to difficulties in shifting attention, according to Baynes, Holtzman and Volpe (1986), which should be borne in mind in the context, below, of holding and shifting attention during information processing. Similarly, studies with normal subjects, in which electroencephalogram (EEG) α desynchronization measures, positron emission tomography, regional cerebral blood flow, and reaction time measures were employed again confirm that while the temporoparietal areas of the LH may be activated by novel stimuli predominantly in right hemispace, the RH is activated by such important events on either side of the midline. However the LH most also have some residual ipsilateral capacity because significant neglect is generally absent after right hemispherectomy and forebrain commissurotomy (Plourde & Sperry, 1984).

So can the LH operate in both halves of extrapersonal space, with—in the presence of intact commissures—differential suppression of left-side awareness by a focally damaged RH? This model suggests that damage to the RH disrupts attentional functions of an entire integrated system, interfering with the expression of compensatory abilities in the intact as well as in the damaged hemisphere, since a damaged and nonfunctioning RH retains its dominant and suppressive role over the LH with respect to attentional functions—a mirror-image version of the conventional account of left hemisphere language dominance. Heilman et al. (1984) demonstrated the role of the commissures in integrating hemisphere–hand and hemisphere–hemispace mapping, which may be dissociated when the arms cross the midline. Thus when a commissurotomy patient tactually bisects a line with the right hand in left hemispace, the disconnection prevents the right hand from being influenced by the RH, which is critical for mediating attentional–intentional functions in left hemispace. The hand will instead be subject to the influence of the LH, which of course also subserves sensory–motor processing of the right hand along the conventional anatomical pathways. The attentional–intentional system of the disconnected LH will direct the sensory–motor apparatus towards contralateral right hemispace and so will generate errors to the right of the true midpoint when the right hand tactually bisects in left hemispace. This is exactly what Heilman et al. found, together with errors to the left of the true midpoint whenever the left hand bisected lines in right hemispace. In both cases, such errors were far more pronounced when the arms crossed the midline, each limb erring towards its own hemispace, a consequence of the two (disconnected) hemispheres each mapping two incongruent (because the arms were crossed) relationships, the hemispatial and the sensory–motor. In normal subjects Heilman et al. suggest that the tendency of each hemisphere to "intend" towards contralateral hemispace may be counteracted by the other hemisphere, an opposition that is lost when one hemisphere is damaged or the commissures are severed. Indeed, we should perhaps note that such observations, together with the finding that patients with hemineglect take longer to move a lever *towards* the neglected left hemispace than further into the unimpaired right hemispace (indicating hemispheric mediation of *intention towards* contralateral space as well as intention and attention within it), are not fully compatible with the idea of a topographical brain map of contralateral space. As we shall see below, the idea of attentionality–intentionality must also be incorporated into such a model. Indeed, the fact that patients can attend to events on the left if they are in effect "bribed" to do so suggests the importance of an additional motivational component.

Line Bisection by Normal Subjects

We have already seen that when patients with left hemineglect try to bisect a line that lies across their midline, they tend to place the transection to the right of the true center (Heilman & Valenstein, 1979); it is as if the area to the left of center is seen as smaller than it really is. In an early study, Rosenberger (1974) investigated visual line bisection in normal subjects. Here, rather than actually transecting the lines, the subjects tried to discriminate between lines that were either accurately bisected or asymmetrically transected. No asymmetries were observed, but this could well have been due to the relative coarseness of Rosenberger's difference scale. Since then, Ramos-Brieva, Olivan, Palomares, and Vela (1984) and Schenkenberg, Bradford, and Ajax (1980) have investigated leftwards errors in line bisection by clinical and control subjects. When we required subjects actually to bisect lines placed across the midline (Bradshaw, Nettleton, Nathan, & Wilson, 1985), we found that they consistently placed the transection slightly to the left of center, a phenomenon that may be labelled left side underestimation (LSU). It is as if such subjects see the extent to the left of center as larger than it really is, possibly because of the greater visuospatial processing power of the right hemisphere (cf. Bradshaw & Nettleton, 1981, 1983 for review), and so in compensation "make" the left side slightly smaller to seem equal to the right. Thus when our subjects transected 10 lines, ranging in length from 80 to 170 mm, they produced a highly significant ($p < .001$) LSU averaging 1.6% of the true half length, an effect shown by 22 of the 24 right-handed subjects. Interestingly, when on another occasion (Bradshaw, Nettleton, Wilson, & Bradshaw, 1987) we required 5-year-old right- and left-handed children to perform the task, using both the preferred and the nonpreferred hand, the right-handed children showed a LSU like that of adults, the value being slightly smaller with the right hand than with the left. Left-handed children, however, placed the transection well to the left with the left hand and far to the right with the right hand; they seemed unwilling even to approach the midline with either hand. We discuss these effects in more detail (below), but should note at this point that a somewhat similar finding has been reported even with left-handed adults by Scarisbrick, Tweedy, and Kuslansky (submitted).

Anatomic Connectivity and Hemispace

It has long been known that a right visual field (RVF) superiority, or a right ear advantage (REA), occurs with lateral presentations of verbal material (Bradshaw & Nettleton, 1983; Bryden, 1982); the opposite applies with certain classes of nonverbal or visuospatial stimuli. In the auditory and tactual modalities, laterality effects have traditionally been ascribed to the prepotency of the contralateral (over the ipsilateral) afferent pathways projecting to a hemisphere specialized for a particular mode of processing. For example, the REA for verbal material is said to arise because of superior connections between the right ear and the verbally specialized left hemisphere (Kimura, 1961, 1967). Under these traditional experimental circumstances, however, receptor–effector location and hemispace may be systematically confounded. Hemispace here refers to the extracorporeal space to left or right of the body midline, within which stimuli may occur and responses may be initiated, as distinct from ear of entry, hand, or visual field. This confounding means that asymmetries in the classical laterality studies may not be the simple consequence of anatomic connectivity. Rather, these differences in performance may be a function of the hemispace (left or right) in which stimuli and/or responses occur. If each hemisphere mediates events that occur in the contralateral hemispace, then improved performance would be expected when the spatial location of the stimulus or response is contralateral to the hemisphere that is more efficient at performing the task. The differential contributions of anatomical and hemispatial factors can only be assessed if receptor–effector location and hemispace are unconfounded.

Kinesthetic Studies

Hand and hemispace can be conveniently unconfounded by employing a design whereby the hands operate in their own (ipsilateral) hemispace, and also cross the midline into the opposite (contralateral) hemispace. Bowers and Heilman (1980) required their blindfolded subjects to explore tactually a rod that lay to the left or right of, or across the body midline. They were to point with the forefinger of the left or right hand, held either in its own hemispace, or across the midline, to the subjective center of the rod. The authors obtained a LSU (when the task was performed with the rod across the midline or in the right hemispace), which they termed a "pseudoneglect," by analogy with the left hemineglect shown by patients suffering from trauma to the RH, though the direction of the two phenomena are, of course, opposite. They also found that performance was more accurate in left hemispace than at the midline or in right hemispace. We (Bradshaw, Nettleton, Nathan, & Wilson, 1983) performed a version of this experiment with left or right hands in their own or opposite hemispaces; subjects tried to subdivide (into halves, quarters, or fifths) a laterally located rod, using only tactual and kinesthetic information. Although we found no evidence of a LSU per se, the left hand significantly ($p < .05$) underestimated relative to the right hand. Moreover, overall performance (i.e., irrespective of directional effects) was slightly more accurate in left than in right hemispace, though the effect just failed to reach statistical significance.

Interestingly, a significant left-hand underestimation relative to the right hand also emerged in another tactual experiment in this series. Subjects were required bimanually to bisect vertical rods, positioned in left or right hemispace with one hand always crossing the midline. (The rod passed vertically through a laterally located horizontal baseplate and one hand kinesthetically measured and adjusted the extent above the baseplate, while the other operated in the same way beneath it.) Subjects wore goggles that occluded lateral vision while allowing central fixation. Subjects next performed a bimanual integration task, with the rod now lying across the midline and the two arms either each occupying their respective hemispaces or crossing the midline. One hand attempted to reproduce on one side an extent or ratio (1:2, 2:3, etc.) of an extent perceived by the other hand on the other side, by laterally moving the rod through a central piece of short pipe. We found that the extent in left hemispace was consistently underestimated by about 1% ($p < .01$). Thus, overall, though neither hand over- or underestimated relative to its fellow, the extent in left hemispace was consistently underestimated.

These effects completely disappeared, however, when the subject performed a version of the task with the head turned 90 degrees to the left or right, thus dissociating the coordinates of head and body hemispace. There was therefore no LSU, either with respect to head or body hemispace, whether the rod ran from side-to-side or front-to-back. (To achieve front-to-back passage of a rod through the subject's midline, the whole experiment was performed upon a specially constructed stool with the rod running directly beneath the seated subject; a pipe extended out either between the subject's legs and behind and beneath the coccyx, or on either side of and below the hips. Subjects laterally adjusted the rod, which passed through and was longer than the pipe, so that the protruding extents at either end were felt to be equal in extent). Thus alignment of the coordinates of head and body hemispace may be necessary for asymmetries to appear.

Moreover, early visual experience may also be necessary for the development of a proper sense of extracorporeal space. In a group of congenitally blind adults tested under normal orientation conditions, LSU was absent when the subjects arms were uncrossed, and reversed (i.e., a left-side overestimation) when the arms were crossed (Bradshaw, Nettleton, Nathan, & Wilson, 1986). In this context it is worth noting that hemispace asymmetries were also absent when long-term blind subjects performed a braille reading task (Bradshaw, Nettleton & Spehr, 1982). "Conventional" asymmetries are also claimed to be reduced in the congenitally blind (Harris, 1980; Hermelin & O'Connor, 1971a, 1971b; Karavatos, Kaprinis, & Tzavaras, 1984; Larsen & Hakonsen, 1983) and the long-term profoundly deaf (Bonvillian, Orlansky & Garland, 1982; Boshoven, McNeil, & Harvey, 1982; Cranney & Ashton, 1982; Gibson & Bryden, 1984; Neville, Kutas, & Schmidt, 1982; Weston & Weinman, 1983). Indeed we shall see (below) that patterns of vibrotactile reaction-time asymmetries may also change in rehabilitated hemineglect patients tested with eyes closed versus eyes open.

In conclusion, there may be an underestimation of extents "perceived" by the RH, either in terms of left *hand* or left *hemispace* performance, the latter being easier to demonstrate when the extents symmetrically cross the subject's midline, rather than exclusively occupying one or other hemispace.

Visual Studies

We have demonstrated (above) the abolition of LSU in a horizontal rod task when head and body hemispace are dissociated by 90 degree head turns to left or right; head and body spatial coordinates must be aligned perhaps for adequate perception of extrapersonal space. We next asked what are the roles of retinal (corporeal) and gravitational coordinates (Bradshaw et al., 1985). These can be dissociated by getting subjects to recline horizontally on one or other side, with stimulus rods horizontal or vertical. Subjects performed a visual rod bisection task; they maintained fixation upon a central ring through which a rod passed, and adjusted the rod extremities (seen in periphereal vision) until they were judged equal. Such an arrangement of course also ensured that the two ends of the rods would project to opposite hemispheres, as scanning was not permitted. A LSU would now incorporate both a hemiretinal–hemispheric and a hemispatial–hemispheric component, and it might therefore be larger than that which we had previously obtained (1.6%) for visual line bisection, when subjects freely scanned the lines before transecting them. In the visual rod bisection task we obtained a very large (3.75%) and significant ($p < .001$) LSU when subjects were upright, an effect that was significantly reduced when subjects lay horizontally, thus dissociating gravitational and retinal coordinates of space.

We can now ask whether the clinical phenomenon of left hemineglect is due to a reduced salience or "attention-getting quality" of events to the left of the midline. Conversely, with normal subjects, are extents to the left excessively salient, perhaps because of the relatively greater visuospatial information-processing capacity of the RH? We (Bradshaw, Nathan, Nettleton, Wilson, & Pierson, 1987) varied the salience of the test rods by manipulating the contrast with the background (left or right sides). Thus we had all black rods, or all white rods, on a horizontal background half of which (to the left or to the right of the subject's midline) was of contrasting tone (black background beneath a white rod, or vice versa); the other half was of nonconcontrasting (i.e., similar, low salience) tone, (white background beneath white rod, or black beneath black). For control purposes we also included a uniform background of high salience: that is, an all white background beneath a black rod or vice versa. The midpoint of the uniform background, or the black–white juxtaposition of the two sharply demarcated and contrasting backgrounds, lay clearly marked exactly in front of the subject's midline. The subject adjusted the rods, by means of wires attached to their ends, so that the two extremities were judged equidistant from the marked midpoint. We included conditions of fixation upon the midpoint, and free scanning, to verify that LSUs are indeed greater under the former condition. We found this to be the case ($p < .025$, 2.65% vs. 1.62%). The fact that the value for free scanning (where only hemispatial factors can operate) was considerably more than one-half that of the value for central fixation (where both hemispatial and hemiretinal factors would operate) suggests that hemispatial factors are more important than the traditional anatomic pathway variables (Kimura, 1961, 1967) associated with hemiretinal factors. When we examined the effect of salience, however, LSUs proved to be biggest when salience was low on the left and high on the right, intermediate with a uniform background, and smallest when salience was low on the right and high on the left ($p < .05$). We must therefore conclude that while a manipulation of salience does affect the LSU in normal subjects, the phenomenon is not due to excessive salience of stimuli to the left of the midline, since a reduction in left-side salience only serves to increase the LSU. Indeed, it is as if fainter stimuli (extents on the low salience side) are somehow seen as bigger; again, perhaps this is because they require more processing resources, and in compensation, are "made" slightly smaller to "appear" to equal the extent on the high salience side. We (Bradshaw, Bradshaw, Nettleton, & Wilson, 1986) addressed these issues by requiring normal subjects to bisect the space between two point sources of light, which were either both constantly illuminated (or both simultaneously flashing) with one light dim and the other bright, or one light on longer than another when simultaneously flashing, or both flashing alternately with same or different durations. We found that faint, brief, or flashing stimuli somehow attracted the subjective midpoint when either hand was used, and that the consistent effect of placing the subjective midpoint to the left of true center was more pronounced when the left hand was used.

We next (unpublished findings) presented subjects with a single vertical target rod in left or right hemispace for 5 seconds. They then adjusted a vertical test rod located in the same or opposite hemispace to equal the remembered length of the target, under conditions of free scanning. We found that a target-right/test-left configuration produced a significantly greater ($p < .01$) LSU than the opposite configuration. Moreover, while left and right target hemispace did not differ from each other, there was a much greater ($p < .001$) underestimation for left (compared with right) test hemispace. This indicates that the phenomenon is not memory dependent (target based), but is instead related to the instantaneous perception of the test item. Furthermore, this experiment suggested that it is truly a case of LSU rather than right side overestimation, as the target-left/test-right configuration did not produce overestimation, only a (nonsignificant) underestimation. We must then conclude that stimuli to the left of the midline are seen as "bigger" than they really are, but not more salient. Again, it is perhaps because the RH has more visuospatial processing capacity that compensation in matching tasks causes extents to the left to be made smaller; as a result both extents appear equal to the subject.

Finger Stimulation, Spacing, and Sequencing Tasks

We have seen (above) that in our first kinesthetic experiment, a magnitude estimation task was performed rather more accurately when it took place entirely in the left hemispace, as compared with the right hemispace; this effect was separate from the general phenomenon of a LSU, but similar to it, was probably a common manifestation of superior spatial processing by the RH. Blindfolded right-handed children (aged 3.5 and 5 years) next had textile patches brushed across the tip of the forefinger of the left or right hand, located in left or right hemispace (Burden, Bradshaw, Nettleton, & Wilson, 1985). This target stimulus was followed by a test patch that was either tactually identical to the target, or differed from it (the level of difficulty of the difference discrimination being individually adjusted for each child). The test patch was administered to the same or opposite hand as the target, held in same or opposite hemispace as before. We found no hand asymmetries whatsoever, but obtained a significant ($p < .025$) left hemispace superiority for the initial target presentation, though not for the subsequent test hemispace, suggesting in this situation that memory rather than perceptual factors may determine hemispace asymmetries. Either hand performed better when located in its own hemispace than when it crossed the midline, an effect that was much greater for young ($p < .025$) and female ($p < .025$) than for older, male children, and it may reflect otherwise established differentials in the rates of commissural maturation (Hewitt, 1962; Rakic & Yakovlev, 1968; Salamy, 1978; Yakovlev & Lecours, 1967).

In another task with an essentially similar hand–hemispace design, blind-folded 5-year-old boys and girls reproduced static configurations of finger spacings that had been moulded on their hands by the experimenter and then scrambled by clenching the fist. This time there were both hand and hemispace asymmetries, again only at the level of the initial (memory) target and not at the level of the subsequent (perceptual) test. Thus performance was superior for the left (versus right) target hemispace ($p < .001$), and for the left (versus right) target hand ($p < .01$). The fact that both hand and hemispace effects emerged may be due to the combination of spatial, kinesthetic, and motor components in this task. Once again, either hand was better ($p < .05$) when operating in its own hemispace, rather than crossing the midline. Finally, in a sequential finger-touching task designed to appeal to the LH, 8-year-old right- and left-handed children were touched, serially, on the fingers of the left or right hand, held in left or right hemispace. They reproduced the target sequence with movements of the fingers of the same or opposite hand held in same or opposite hemispace. Right-handed children showed no hemispace asymmetries, though both the right target and the right test hand were superior to the left (in both cases, $p < .025$), and again either hand performed better in its own hemispace. With left-handed children there were no hand asymmetries, and either (target) hand was better when in right than in left (target) hemispace. We can conclude that for this task hemisphere–hand connections predominate in right-handed children, and hemisphere–hemispace connections predominate with left-handed children, who otherwise might have been subjected to a dissociation between the preferred hand and the mediating hemisphere. In either case a superiority of the LH emerged in this active sequential task, with again effects appearing stronger at the level of initial (memorized) target than for the subsequent (perceptual) test stimulus.

Vibrotactile Experiments

So far we have measured performance accuracy rather than speed while investigating hemispace asymmetries. Reaction times (RTs) to vibrotactile stimuli are a convenient way of unconfounding the effects of anatomical connectivity and hemispace while using a speed measure. In our first such experiment (Bradshaw, Nathan, Nettleton, Pierson, & Wilson, 1983), dextral subjects, who fixated ahead pressed a button with the thumb when a transducer on the adjacent index finger was felt to vibrate. The two hands were tested separately and were positioned in their own or in contralateral (i.e., across the midline) hemispace, for a block of trials. Twenty-two out of 24 dextral subjects gave a highly significant ($p < .001$) right hemispace superiority of 9 ms, the hands not differing. We then unconfounded head and body hemispace by requiring subjects to turn the head 90 degrees to left or right, with the stimulated and responding limb held out either in the midline, or to the side of the body. All hemispace asymmetries were lost with respect to the body (hand held out from side) and greatly reduced with respect to the head (hand held in the midline). So just as 90-degree head turns earlier destroyed LSUs in a rod bisection task, so also right-of-body hemispace superiorities were lost and right-of-head hemispace superiorities greatly reduced under similar circumstances in a vibrotactile RT task, which dissociated the head and body components.

What happens if instead we dissociate gravitational and corporeal coordinates, as in the visual rod bisection task, by getting subjects to perform the task while recumbent on one or other side? By corporeal we mean left and right with respect to the spinal axis, irrespective of posture; by gravitational left–right we mean with respect to the gravitational vertical, while the now recumbent subject is still facing in the direction in which he or she is looking. Thus when a recumbent subject lies on the left side, gravitational left extends out beyond the top of the head, and gravitational right towards the knees. These relationships reverse when the subject lies on the right side. We repeated our first vibrotactile experiment with blocks of trials for hand (left, right), posture (lie on left, right) arm (up, down, or beyond head, between knees), and analyzed the data in terms both of gravitational and corporeal coordinates (Bradshaw & Pierson, 1985). There were no significant hand or hemispace effects whatsoever; dissociation of gravitational and corporeal hemispace, which led to a significant reduction of the LSU in visual rod bisection, also appears to destroy vibrotactile hemispace asymmetries.

Are right side advantages (RSAs) for vibrotactile stimuli motor or sensory? If one hand is stimulated and the other responds, we can separately assess the independent contributions of left and right sensory hand, left and right motor hand, left and right sensory hemispace and left and right motor hemispace. Under these circumstances we found that only motor hemispace gave a significant ($p < .01$) RSA (6 ms). So the locus of this effect appears to reside in the preparation of a response, rather than the discrimination of a *signal*. This effect was replicated in a later study with clearly perceptible stimuli (a 4-ms motor RSA; $p < .01$); but when stimulus intensity was reduced to a near threshold level the motor RSA was lost, and a hand asymmetry appeared. Thus right-hand-responding, left–stimulated was 9 ms faster than the opposite configuration ($p < .025$). However in this low–intensity condition there was also a 4 ms sensory RSA that just failed to reach significance, although it was shown by 17 out of 24 subjects. The relative attentional demands of the sensory and motor components of the task seem to determine the nature of vibrotactile asymmetries. With high–stimulus intensity, attention can be allocated to the motor component, and it is at this level that asymmetries appear. When stimulus intensity is reduced, however, attention must be directed to the sensory aspects of the task and there is a tendency for performance to be dependent on the location of the sensory hand.

Hitherto, we have obtained vibrotactile RSAs with trials to one or other side presented in blocks, rather than randomly. Does the phenomeon depend upon the establishment of position sets, dextral subjects having a greater ability to hold covert attention to the right (compared with the left) for a block of trials? If this is the case, then randomization of the side of stimulus and response should lead to a loss of effects. Subjects must now divide or switch attention between the two sides of space if they want to obtain maximal performance. If attention is therefore directed to both sides of space, then responses made on the right cannot benefit from a greater ability to focus attention in this region of space for an extended period.

We adopted just such a randomly alternating paradigm, each hand having its own stimulus and response unit, the stimulated hand being the one to respond (i.e., a choice RT task). Hands were held either in their own (uncrossed) or opposite (crossed) hemispace. We found that there was no difference between response latencies in left and right hemispace, which suggests that the vibrotactile RSA is attentional in origin. In this task, however, there was a significant ($p < .025$) right hand superiority (13 ms), and the crossed configuration was now for the first time considerably slower than the uncrossed ($p < .001$). Of course this constitutes a hand/hemispace spatial–compatibility effect—each hand is faster when positioned in its own hemispace. Spatial compatibility effects are of course traditionally absent from simple RT tasks, and present in choice RT situations (Bradshaw & Umilta, 1984); conversely, hemispace effects may be absent from choice tasks and present in simple RT tasks, where the spatial coding of limb position is unimportant. If so, we can ask whether hemispace asymmetries reappear, at the expense of hand asymmetries, if we employ blocks of trials for a given responding hand or side, but randomly alternate, as before,the stimulated hand or side. As the vibrotactile RSA has been shown to be motor rather than sensory, a RSA might be predicted when side of stimulation is blocked. When we performed just such an experiment there was once again no hemispace asymmetry, but the right *hand* was 8 ms faster ($p = .05$) than the left. Perhaps not surprisingly, when the same hand was stimulated and responded, performance was also faster (18 ms; $p < .05$) than when opposite hands were stimulated and responded. Moreover there was now no longer any advantage for the uncrossed configuration, so the need to select a response hand may determine whether or not there will be hand/hemispace spatial compatibility effects. Hemispace asymmetries may only occur when attention can be focussed to one or other side for a block of trials and wholly allocated to the motor aspects of the task.

We can conclude from these vibrotactile experiments that hemispace asymmetries will usually occur only when head and body coordinates and corporeal and gravitational coordinates are aligned. Moreover position of an event in extracorporeal space is at least as important as traditional anatomical connectivities (Kimura, 1961, 1967) in determining asymmetries. Vibrotactile RSAs occur with simple, not with choice RTs: that is, where spatial position is unimportant in stimulus–response coding, and only when the stimulus is clearly perceptible and trials are presented in blocks to one or other side. They may reflect a greater ability to hold covert attention to the right than to the left. However, our view about the role of attention is not compatible with Kinsbourne's (1973) attentional model: he invokes an activational component, whereas our account implicates sustained attentional effects. He predicts that asymmetries should occur only when stimuli and responses occur randomly as to side, rather than when side of stimulation and response is blocked. We, of course, find that hemispace asymmetries are absent when side of stimulation, or stimulation and response is random and present when lateral stimulation occurs in blocks.

In a recent vibrotactile experiment (Pierson-Savage & Bradshaw, in press) we mapped RTs at a number of positions in the space in front of the body. The stimulated and responding hand (left or right) was placed out from the body, either at the midline (0 degrees) or at 15 degrees, 45 degrees or 90 degrees to the left or right of the midline (i.e., at seven possible positions). Reaction times for the two hands were identical, confirming the absence of hand effects when there is no position uncertainty, and there was no hand by position interaction. There was, however, a significant effect of position. Performance was found to be relatively uniform in the region from 90 degrees right round to 45 degrees left, but was markedly slower at 90 degrees left. The finding that responses in most of left hemispace were as fast as those in right hemispace leads to the conclusion that our vibrotactile RSA should perhaps instead be characterized as a disadvantage for responses made by either hand in the far left part of left hemispace.

It is possible that this disadvantaged region may extend behind the body, although this remains to be determined. Moreover the findings suggest that hemispace effects, at least in the vibrotactile RT paradigm, may not perhaps be a direct consequence of superior left hemisphere mediation of the task, coupled with a functional relationship between each hemisphere and the contralateral half of space. Rather they should perhaps be viewed as reflecting differences in the ability to focus covert attention as a function of spatial location; right-handed persons may have difficulty in focusing attention for a block of trials, in the far left part of the space in front of the body. This difficulty may perhaps arise because people who are strongly dextral perform most manual tasks which require *sustained* attention in the region of space extending from about 45 degrees left to (at least) 90 degrees right. This area could be said to represent a "functional processing space." However, very few tasks of this kind are carried out in the far left part of space (beyond about 45 degrees left). Consequently, dextrals may find it more difficult to focus attention in this region than in the functional processing space.

The specific mechanisms underlying this apparent difficulty in holding attention to the left remain to be determined. It may simply be due to a lack of experience in focussing attention in far-left space, and may therefore be overcome with extended practice in attending to events in this location. Alternatively, the difficulty could be a consequence of a right bias that has developed because, in right-handed persons, attention is usually focussed for extended periods only within the functional processing space, and practice might not overcome this bias. Whatever the mechanism involved, the results of our vibrotactile experiments seem compatible with the view that differences in vibrotactile reaction time as a function of spatial position are attention in origin.

Auditory Hemispace

The most direct test between Kimura's anatomic-pathway model and the hemispace approach lies in the auditory modality. She specifically claims (1961, 1967) that the dichotic right ear advantage (REA) is due to the stronger, dominant, contralateral auditory pathways suppressing the weaker ipsilateral route, so that the right ear has priviledged access to the verbal LH. We (Pierson, Bradshaw, & Nettleton, 1983) replaced the traditional earphones (which separately stimulate each ear) with a single, laterally located, loudspeaker. Through this we played competing unilateral stimulation, sequences of aligned pairs of a digit (1, 8, 9, or 10) and a letter (g, v, k, or z), which were not mutually confusable, one spoken by a male and one by a female as a further aid to discriminability. The stimuli, played from the single loudspeaker on one or other side of the subject, were loud enough to stimulate both ears. Subjects (right-handed males, audiometrically screened to eliminate threshold differences between ears) shadowed (i.e., immediately repeated) each stimulus pair, and the vocal naming latencies were measured. There was a RSA of 38 ms ($p < .005$) which was demonstrated by 10 out of 12 subjects, the two reversals being very small. This finding conclusively demonstrates that competition at the two ears, as with dichotic stimulation, is unnecessary for demonstrating a verbal REA, and that Kimura's structural model is inadequate.

We next asked whether it is the real or the perceived direction of a sound source that determines RSAs. We sought to generate "pseudo" RSAs via a ventriloquistic technique that relies upon the phenomenon of visual capture of the apparent direction of an ambiguous auditory source. We placed two operating loud speakers one directly in front of, and one behind the subject's midline, one voice from the previous tape being channelled to one loudspeaker and the other voice to the other. This arrangement created considerable positional uncertainty about the direction of a perceptually composite signal. An inactive dummy loudspeaker was placed either on the left or the right of the subject, who believed it to be fully operational. The dummy speaker effectively captured the sound source, creating a powerful directional illusion. Eleven out of 12 subjects were faster ($p <$ .005) by 28 ms when the dummy loudspeaker was on the right rather than the left side, indicating that it is the perceived, rather than the real position of a sound source that determines hemispace asymmetries. We next, as in our rod bisection and vibrotactile RT tasks, sought to dissociate the two coordinates of (auditory) hemispace, head and body, by 90 degree head turns to left or right. When an (active) loudspeaker was placed before or behind the subject's midline, it lay to the left or right of the turned head. (We were careful to ensure that the only subjects used were able to maintain the requisite full 90 degree head turns). Likewise when the loudspeaker was to left or right of the body, it lay exactly before or behind the head midline. Under these circumstances all hemispace asymmetries again completely vanished, and RTs were identical for the averages of the head-turned-right and head-turned-left conditions, itself a finding that counts against Kinsbourne's (1973) activational–attentional account. (According to him, performance in a verbal task should be superior when the subject orientates to the right). Finally, we confirmed that the loss of a RSA with 90 degree head turns is a robust phenomenon by playing competing unilateral stimulation over a proximal earphone, rather than a distal *loudspeaker* source. We used experienced subjects from the previous experiment in this series and the same stimulus tape. As a control a head front (no turn) condition was reincorporated. We found that while the *overall* RTs for head front, right and left did not differ significantly, with head front there was a very large (64 ms) and significant ($p < .001$) REA, and no significant ear asymmetries for either of the two turned-head conditions. Finally (unpublished data) we asked our subjects to lie horizontally upon one or other side to dissociate gravitational and corporeal coordinates of space, just as we had done in the visual rod bisection and vibrotactile tasks. We used the earphone version of the task and once again there were no asymmetries.

Overview and Conclusions

It is well established that patients suffering from RH damage and hemineglect typically bisect a visually presented horizontal line to the right of the true midpoint, as if a large extent to the left of center is seen as smaller than it really is. Our recent work with normal subjects has demonstrated an analogous and possibly related phenomenon. Right-handed subjects consistently bisected lines slightly to the left of the true midpoint, as if the extent to the left of center is seen as larger than it really is. This LSU has also been demonstrated in our own and other recent studies where rod bisection tasks were used. The LSU is larger when the line or rod crosses the midline rather than being positioned wholly in left or right hemispace, or when it is horizontally rather than vertically oriented. Although "conventional" lateral asymmetries are reported to be stronger when the task involves memory (Moscovitch, 1979; Nettleton & Bradshaw, 1983) the LSU is immediate and perceptual, rather than the product of a fading memory trace. The phenomenon is not purely sensory however; attentional components (as we will shortly argue) also contribute to the effect.

It seems probable that the effect is a true LSU, rather than a case of right-side overestimation. There are, however, methodological difficulties associated with any attempt to unconfound experimentally these two possible determinants of the phenomenon. Any experiment that aims to do this must overcome the problem of a central standard or target extent being seen as lying to the left of a variable comparison extent on the right, or to the right of a comparison extent on the left.

The asymmetric matching of extents in the two hemispaces may be accounted for in terms of differences in the visuospatial information–processing capacities of the two hemispheres. This difference may itself be a consequence of language preempting visuospatial processing space in the LH. If there is a functional relationship between left hemispace and the RH, then extents to the left may be seen as larger because the RH has relatively greater visuospatial processing power.

Though leftward extents may be seen as larger, it seems that the LSU is not the result of a greater salience of stimuli to the left side. When we differentially manipulated the contrast between the test rod and backgrounds on the left and right sides, we found that the LSU is greatest with low salience on the left side and high salience on the right. If the perceived magnitude of an extent is somehow determined by the amount of processing capacity (i.e., attention) that it is allocated, then low-salient, poorly contrasting stimuli might appear larger because they require an increase in processing capacity.

If the LSU is the result of hemispheric differences in visuospatial processing ability, it is then not clear why the choice of hand used to perform the task can greatly influence the magnitude and direction of visuospatial asymmetries of this kind. We found that 5-year-old, right-handed children showed larger line bisection LSUs when the task was performed with the left hand than when it was performed with the right. Similar findings have been reported by Ramos-Brieva, Olivan, Palomares, and Vela (1984) and Schenkenberg, Bradford, and Ajax (1980). Both groups of researchers made the incidental observation that adults gave a larger LSU when lines were bisected with the left hand. The 5-year-old, left-handed children in our study behaved rather differently, placing the transection far to the right when using the right hand and far to the left with the left hand. Indeed, Heilman, Bowers, and Watson (1983) report that they observed the same pattern of errors when a patient with a callosal lesion was asked to point to her midline. Heilman, Bowers, and Watson (1984) have also reported that in tactual and visual line bisection tasks, the hands of commissurotomy patients, when crossing the body midline, err towards their own hemispace. Is it possible then, that the behavior of the young sinistral children and the patients may in fact be causally related? Rakic and Yakovlev (1968) report that the commissures are not completely mature until the end of the first decade, and it could be that commissural maturation is delayed in left-handed as compared with right-handed children. The 5-year-old left-handed children we studied may therefore be functionally acallosal and this may explain, at least in part, their performance in the line bisection task. Indeed, we note that callosal morphology is now known to differ between right-handed and left-handed adults (Witelson, 1985), and according to Scarisbrick, Tweedy, and Kuslansky (submitted) a greater between-hand separation in the locus of line bisection may be a characteristic even of adults who are left handed. It should be noted, however that in a just-completed study (Bradshaw, Spataro, Harris, Nettleton, & Bradshaw, in press), while we confirmed the occurrence of a kind of symmetrical neglect in the line bisections of 4-year-old children, and while left-handed children from ages 4 to 8 years continued to show bigger between-hand separations than left-handed children, the pattern of symmetrical neglect did not on this occasion appear to be peculiar to very young left-handers as opposed to right-handers of the same age. Moreover in a timed peg-moving task, young left-handed children did not slow when crossing the midline, or perform better with centrifugal abductive movements, findings that are not easily compatible with the idea of callosal immaturity or anomalous visuomotor organization in young sinistrals.

Though right-handed subjects show larger line bisection LSUs when the left hand is used, the converse is true for visual rod bisection tasks. Right-handed adults produce larger LSUs when a laterally located rod is adjusted by the right hand. There are, however, some key differences between the two tasks that may underlie these seemingly discrepant observations. The former (line bisection) involves bisection of centrally located stimuli, while in the latter task the stimuli are peripherally located and a magnitude match is required. Otherwise performance in both cases is influenced by the hand used. In a task involving vision alone, we have found, informally, that when two extents of equal magnitude appear side by side they are judged to be identical. The absence of an LSU when the task lacks a motor component supports the claim of Le Doux, Wilson, and Gazzaniga (1977) that asymmetries are generally more pronounced at a *manipulospatial* than at a purely perceptual level. Indeed the role of motor factors becomes further evident in an unpublished study of ours whereby right-handed subjects unimanually matched extents on either side of the midline by feel; we found that extents in the hemispace contralateral to the one normally occupied by the responding limb were set very much smaller than those in the limb's own ipsilateral hemispace, though less so when adjustments were made by the left hand rather than the right hand. This combination of greater accuracy with the left hand, and a reduction of magnitudes in contralateral hemispace, does of çourse lead to smaller settings in left hemispace, though the comparatively small size of these resultant asymmetries warns us agaisnt invoking this explanation for bimanual and visual situations generally.

These recent experiments involving rod and line bisection have some interesting comparisons with early reported work examining magnitude estimation in the two halves of the visual field (see Ritter, 1917). A number of variables including body posture were investigated in these early experiments and some researchers even used terms such as left-side underestimation, but in a sense that is opposed to ours. We mean that an extent on the left is set smaller than it really is, possibly because the greater visuospatial processing capacity of the RH results in a perceptual magnification of leftward extents. However, others in the past have meant that an extent was seen as smaller than it really was, and in compensation was therefore made larger. These early studies did not however reveal consistent trends across individuals and until recently the issue has been almost completely ignored (though see Walker, 1972). However, our recent findings have now confirmed that in rod and line bisection tasks, there is a consistent LSU both within and across individuals.

We suggested at the beginning of this section that the slight LSU shown by normal subjects might be related to the left side overestimation that typifies the performance, on a line bisection task, of patients with damage to the right hemisphere and hemineglect. Do similar mechanisms underlie the lateral errors made by the two groups? In some preliminary work aimed at addressing this question we have tested a number of patients on the tasks used in our studies with normal subjects. One dextral patient we examined (N.M.) had suffered a stroke and sustained extensive RH damage 2 years previously. He was diagnosed as having severe hemineglect and underwent extensive rehabilitatory training. His performance on our tactual rod bisection task resembled that of our normal subjects: he showed a consistent LSU. However, he seemed actively to be directing his attention to the left, as shown by eye movements that we monitored, even though the task was purely tactual. Further evidence of a leftward attentional bias was revealed by the vibrotactile RT task. Here N.M. showed a left side advantage (LSA) rather than the RSA that is almost always shown by normal right-handed subjects. His performance also contrasts sharply with that of unrehabilitated patients with hemineglect on this task (see below) who typically show very large RSAs, much larger than those of normal subjects.

When we manipulated the salience (contrast of the test rod and the left/right background) in the visual rod bisection task, it became clear that his leftward attentional bias was a strategy that he had learned to use to overcome his hemineglect. When he was given the task with low salience (contrast) on the right side, he overestimated the left extent, that is he gave a typical hemineglect response. It seems that the low salience on the right side of the display forced him to direct attention away from the left and toward the right, thus unmasking a latent left hemineglect. These findings suggest that left hemineglect is not a consequence of stimuli on the left appearing less salient (a low-level sensory explanation). Rather, patients with left hemineglect may indeed suffer a deficit in the capacity to direct attention to the left.

The new procedures that we have developed in our experiments with normal subjects should now allow us to quantify the various stages of the hemineglect syndrome, and the vibrotactile paradigm seems particularly suitable for this purpose. Separation of the sensory and motor components of the task should also allow us to determine the relative contributions of posterior (sensory) and anterior (motor) processes. Of course, this sensory–motor unconfounding requires the use of both hands and will only be possible for patients who do not suffer from left paralysis or paresis, which is commonly found in conjunction with left hemineglect. One patient we have tested had minimal left-sided paresis and was able to do the bimanual version of the vibrotactile task. He showed a sensory RSA of 89 ms and a motor RSA of only 15 ms. With eight patients with hemineglect who could only use the right hand and who had undergone little or no rehabilitative training, vibrotactile responses made on the left side were slower (by 29 ms or more) than responses made on the right. This difference is much larger than the 9 ms effect shown by normal right-handers. The same task was also used to test four rehabilitated patients, who had suffered strokes within the 6 months prior to the test session, but were deemed to have recovered from hemineglect. We found that left-side responses were faster than right-side responses in three of the four patients tested. However, three of these rehabilitated patients have also performed the task with their eyes closed, and all three showed a left side *disadvantage* of 20 ms or more. Thus, with their eyes closed, they behaved like the nonrehabilitated patients. This finding suggests that the left side advantage shown by rehabilitated patients in the eyes-open, central-fixation condition, is a consequence of visually based compensation strategies. Mesulam (1985) has also noted that performance asymmetries in hemineglect patients may alter when they engage in tactual tasks with their eyes closed, and Chedru (1976) has found that hemineglect patients (without visual field defects) demonstrated left hemineglect on a tactual space-exploration task only when their eyes were closed. Indeed according to Blum (1985), the status of the eyes (open or closed) affects the electrophysiological representation of tactual space in the inferior parietal lobule.

Other experiments should allow us to investigate attentional and unilateral memory deficits, stimulus salience, extinction, midline shifts, and distortions of apparent size—all of which may differentially characterize a nonunitary syndrome. For example, the following procedure could be used to investigate the notion that some manifestations of hemineglect (such as a left-side overestimation of extents that cross the midline) may reflect a shift in perceived midline location. The patient could be asked to adjust a horizontally oriented rod so that its two ends are equidistant from a central ring through which it passes. The central ring could be placed either at the patient's true midline or where he or she says it is (to the right of the true midline for a patient with left hemineglect). It performance on the bisection tasks reflects a midline shift, then left-side overestimation would be expected in the former case, with objective equality of the two extents in the latter case.

To return to our work with normal subjects, we found left hemispace superiorities for kinesthetic judgements of extent (almost significant), texture matching, and finger spacing. These effects seemed to appear, or to be stronger, when the task involved memory rather than purely perceptual factors. There was a right hemispace superiority for sequential finger stimulation tasks in left-handed subjects, and a right hand superiority in right-handed subjects. Thus, though both groups employed the left hemisphere, the asymmetry was manifested by hand effects in right-handed subjects, and by hemispace effects in left-handed subjects who, of course, normally employ the left hand for such manipulations and who may therefore otherwise suffer a dissociation between the preferred hand and the mediating hemisphere. A RSA was also found with competing auditory stimulation in a verbal shadowing task. This latter effect depended as much on the perceived (but illusory) position of a sound source as upon its real location. In the vibrotactile RT task, an apparent RSA was demonstrated when there was no uncertainty about the presence or spatial location of the stimulus. A subsequent experiment, however, which measured RTs at several positions in front of the body, revealed that the asymmetry in this task is more correctly characterized as a disadvantage for the far left hemispace. Moreover, this disadvantage may be attentional in origin.

In general, hemispace asymmetries seem to be different from and stronger than anatomic pathway effects. The conventional anatomic connectivity account (Kimura, 1961, 1967) cannot of course accommodate our auditory findings. Similarly, our tactual and vibrotactile findings cannot be explained in terms of superior contralateral connectivity between a receptor or effector and a hemisphere specialized for a particular mode of processing. The experiments in which hemispace effects have been demonstrated indicate that it is the real or perceived position of a stimulus in extracorporeal space that may determine asymmetries, rather than just the proximal locus of receptor stimulation.

The notion that the apparent spatial location of stimuli may determine subsequent perception has also been discussed in a similar context by Deutsch (1980). She describes her octave illusion, whereby two tones, one octave apart, are repeatedly alternated between ears, the initial sequence being simultaneously presented, exactly out of phase, to the two ears. Instead of hearing a simple continuous two-tone chord, whose ear of input for each component is repeatedly switched, the listener typically reports hearing a single tone, switching from ear to ear, whose pitch alternates from high to low. With right-handed subjects, the high tone is usually localized in the right ear, and the low tone in the left. More importantly, exactly similar effects occur if loudspeakers are substituted for earphones, indicating, she argues, that the illusion occurs not along pathways conveying ear information, but with respect to the representation of the two sides of auditory space. Indeed, she holds that what is heard is determined partly by where (in space) the signals seem to come from.

The relatively greater strength of hemispace (than ear-of-entry) asymmetries was shown in our auditory experiments. With relatively inexperienced subjects, the left–right differences were larger with a unilateral loudspeaker (whose sound tends to be localized "out" in space) than with a unilateral earphone (where the locus of the sound tends to be perceived as within the ear itself). Similarly, in the visual rod tasks, the effect for free scanning (hemispatial factors alone) was considerably more than one-half the size of the effect for central fixation (hemispatial plus hemiretinal factors). Finally, in the static finger-configuration task with children, the left hemispace superiority was more reliable than the left hand superiority.

Hemispace asymmetries were found to be lost or significantly reduced when the head was turned 90 degrees to the left or right, resulting in the dissociation of head and body coordinates. The effects were also lost when subjects lay on one or other side, thereby dissociating corporeal and gravitational coordinates. The brain of course does not merely register events impinging upon the proximal receptor surfaces; for good reasons (cf. Gardner & Ward, 1985) it also maps events occurring out in space beyond the body, before such events are directly encountered. Thus the auditory cortex maps the contralateral sound field (Phillips & Gates, 1982), a situation that is not restricted only to species like the bat where this is obviously advantageous (Calford, Grayden, Huerta, Koas & Pettigrew, 1985).

Polysensory areas in the tectum (Harris, 1986), inferior parietal lobule, and possibly even in the cerebellum (Stein, 1986) are involved. The auditory map in the superior colliculus is affected by gaze direction (Jay & Sparks, 1984): the superior colliculus controls eye, pinna, and head orientation (Meredith & Stein, 1985), and may be where the spatial maps from sight and hearing are combined (Meredith & Stein, 1986). In the cortex, eye and head position (including tilt) alter the firing of retinotopic cells (Andersen, Essick, & Siegel, 1985; Reinis, Landolt, Money, Lahue, & Weiss, 1986). Lateral movements of the hands as well as the eyes affect the EEG (Autret, Auvert, Laffont, & Larmande, 1985), while the tactual representation of space may be affected by whether the eyes are open or closed (Blum, 1985). Bisiach, Capitani, and Porta (1985) have observed that, in patients with hemineglect, both gaze direction and trunk orientation influence the boundary of the neglected region of space. These findings raise the possibility that a 90 degree head turn to the right or left might modulate the moment-to-moment symptoms of hemineglect. A similar effect might be expected if the patient was to lie on one or other side. If these suggestions can be verified experimentally, they may open new avenues for rehabilitation of patients with hemineglect.

The above observations suggest that both the perceived spatial location of a stimulus and the perception of one's own position in space results from the integration of complex multimodal inputs (in humans such integration is usually regarded as being a function of the parietal association areas). The brain probably maintains two simultaneous maps, one of which represents the proximal receptor surfaces and mediates conventional lateral asymmetries. The second map may be for stimuli and responses that occur out in extracorporeal space and this representation possibly mediates at least some of the hemispace asymmetries (see also Rizzolatti, Gentilucci and Matelli, 1985). This system of spatial coordinates appears to be altered when there is conflict between the senses, as there was between vision and audition in our ventriloquism experiment where a dummy lateral loudspeaker was used.

There are other ways whereby we can dissociate spatial–cordinate systems, induce conflict between the senses, or mislead the subject that may have interesting empirical consequences. One such visual experiment could involve the wearing of laterally displacing prisms. Weiner, Hallet, and Funkenstein (1983) report that a subject who has worn such prisms until adaptation is complete may, on their removal, sense that a laterally displaced limb is central to the midline or vice versa. Such misperceptions might be expected to affect vibrotactile RT asymmetries or pointing accuracy. Heilman, Bowers, and Watson (1983) asked subjects who had not been wearing prisms to point to an imaginary position in space directly in front of their chest midline. They reported that normal subjects, with their eyes closed and using their right hand, consistently pointed slightly to the left of the true midline. This may be somewhat analogous to the normal LSU in line bisection. In an incomplete study, however, we have found a different pattern of errors. In our task, subjects had their eyes open but were unable to see their hands or arms. When errors were averaged for the two hands, 12 out of 16 right-handed subjects pointed to the right of the true midline; of the four remaining subjects, one showed no overall error and three pointed slightly to the left of the midline.

Hartmann (1983) found that subjects trying to locate ambiguous sound sources were biased to the left of the correct azimuth, the average error being 0.75 degrees. He also found that bias effects increased enormously if subjects shifted their gaze. Once again this latter finding suggests that the auditory map of space is linked to eye position. A wealth of possible research is also revealed in terms of accuracy of manual pointing to visual targets as a function of head and/or gaze direction with respect to each other and/or the body. It would also be interesting to investigate the effect of variation in gaze direction in the vibrotactile and (competing monaural) auditory paradigms. Hartmann (1983) also reviewed evidence of superior localization of sounds in left rather than right space, a finding we ourselves would have predicted. Duhamel, Pinek, and Brouchon (1986) have also reported that subjects were more accurate when pointing to unseen auditory targets located in left as opposed to right hemispace.

Hemispace asymmetries in the visual modality still require systematic investigation. However, there is some evidence that the hemispace (rather than just the visual hemifield) in which stimuli are located may in fact determine asymmetries. Tressoldi (1987) found that lateral asymmetries may appear in visual tasks (involving word and nonword naming or reaction time to faces) even with foveal presentation of stimuli in the left or right hemispace. In this experiment, the head faced forward but eye fixation was deviated to the left or right (cf. Bowers, Heilman, & Van Den Abell, 1981). Other earlier studies have also suggested that there are visual hemispace asymmetries, although the results were not interpreted in this context. Gilbert and Bakan (1973) produced left–left and right–right composite photographs, each derived from one or the other half of an original target face. When all three photographs (the original and the left and right composities) were presented simultaneously, most subjects judged that the left–left composite (derived from the left half of the target photograph) resembled the latter more closely than the right–right composite. Because the task employed free scanning, this finding would suggest that the side of the face in left space receives preferential processing by the right hemisphere, according to a hemispace interpretation. In a study by Foley (1975) targets were positioned at varying distances to the left or right of the subject, who was required to look at and point to the target. Foley reported that angular error (left-right mislocalization of the target) was significantly smaller in the left field (which in this case corresponds to the left hemispace) than in the right field (right hemispace). However, there was also a small, but significant, tendency for overreaching errors to be larger for targets on the left side than on the right. The former result, though, seems analogous to the auditory findings (above) reviewed by Hartmann (1983) and reported by Duhamel et al. (1986). Both findings may also be related to our demonstration of a left side superiority in certain tactual tasks.

It has already been noted that hand rather than hemispace asymmetries have appeared in our kinesthetic tasks involving judgements of extent. Similarly, hand differences (without hemispace differences) have been demonstrated in our vibrotactile RT tasks, but only when the signal intensity was low or when the location of the stimulus and response, or the stimulus alone, was pseudorandomly alternated between the left and right sides. The appearance of these hand effects in the vibrotactile task is rather puzzling and at present they seem to lack a satisfactory explanation. The loss of hemispace effects with low stimulus intensity and with the random-as-to-side design, suggests that these asymmetries only appear when attention can be allocated to the motor aspects of the task and directed in advance to one or other side for a block of trials. They may therefore reflect a difficulty for right-handed subjects in focusing attention for a block of trials, in the far left part of space.

It should now be asked whether an account of asymmetries invoking hemisphere–hemispace mapping (which seems to have partly replaced the anatomic connectivity model) should itself be replaced by one invoking differential attentional biases toward various regions of space. In the vibrotactile modality, right-handed subjects appear more ready to initiate a response in the region to the right of the midline and as far to the left as the hands usually operate in normal psychomotor activity, as long as all sensory and motor aspects of the task are predictable. We have already noted that in this case the disadvantaged region lies far to the left. In the auditory modality, where the RSA seems to reflect left hemisphere language lateralization, it is as if the subject is more ready to respond to stimuli perceived as (really or apparently) coming from the right.

Attentional factors may underlie the loss of vibrotactile hemispace asymmetries when the subject cannot predict the side of stimulation or response. It is possible however that this finding could be at least partly explained in terms of hemisphere–hemispace relationships, if it is assumed that sensory and motor events must be position-tagged ahead of time in order for hemispace effects to be demonstrated. If such events can be tagged ahead of time as occurring in the left or right side of space, then they may be processed by the contralateral hemisphere. In the case of auditory and vibrotactile tasks, stimuli tagged as coming from the right side would therefore be processed by the left hemisphere, which may be more efficient at mediating these tasks. When position information is not available before the event, however, there is no division of labor between the hemispheres. Both hemispheres now begin stimulus processing, but sensory or motor events, regardless of their spatial location, are now mediated by the specialized hemisphere that is always first to complete the required processing. There is therefore nothing to generate an asymmetry in this case. We should of course be wary of assuming that similar mechanisms underlie the vibrotactile and the auditory RSA. A preliminary study, aimed at investigating the relationship between the two tasks, has failed to reveal any correlation between the magnitude and direction of the auditory and the vibrotactile asymmetries across a sample of left- and right-handed subjects. It could be that the auditory phenomenon reflects both LH language lateralization (a true RSA) and attentional processes (a far left side disadvantage). The vibrotactile phenomenon, however, may only reflect the latter, a differential ability to allocate attentional processes (perhaps predominantly the motor processes) around the azimuth of extracorporeal space.

The demonstration that the disadvantaged region for vibrotactile RT includes only the far left portion of left hemispace is of course rather difficult to explain in terms of hemisphere–hemispace relationships. If the left hemisphere is more efficient at performing the vibrotactile task and the hemispheres mediate the two halves of space, it would be expected that performance in the right half of space would be better than performance in the left half of space. This was not the case: performance in most of left hemispace was just as good as it was in right hemispace. It would seem that a conventional hemispace account of these findings would need to claim that in this kind of left-hemisphere-driven task, control by the LH is not limited just to the right hemispace, but also extends some distance around into the left hemispace. Indeed, this seems compatible with the notion (above) that the RH can mediate (to some extent) attention for both sides of space, even though its dominant tendency is toward contralateral (left) hemispace.

This sort of ad hoc explanation, however, is not entirely satisfactory. Moreover the assumption that the hemispheres differ in their ability to perform the task is a necessary component of any "hemispace" account of laterality effects. It is not clear, however, that the LH would necessarily be more efficient in mediating the button-press response that is required in the vibrotactile paradigm. (We have previously shown that it is speed of response initiation, rather than stimulus detection that varies as a function of spatial position; see above.) Though there is some evident (Bradshaw & Nettleton, 1983; Kimura, 1976, 1977) that the left hemisphere may be specialized for complex motor tasks, the vibrotactile response is rather simple and the two hemispheres may not differ in their ability to mediate it. Any attempt to account for the results of the vibrotactile experiments in terms of hemisphere–"hemispace" relationships is therefore problematic and it seems that any account of hemispace effects needs somehow to incorporate the role of attention.

This does not however mean that we can equate our view of attention with that proposed by Kinsbourne (1970, 1975). Briefly, his model has an activational component, whereas our account invokes sustained attentional effects. Kinsbourne's model predicts that asymmetries will be largest when subjects cannot predict which side will next be stimulated, but we of course find that asymmetries are lost under these circumstances. Kinsbourne also predicts larger verbal RSAs with rightward head turns, while our experiments have shown that asymmetries are lost when the head is turned 90 degrees to the right or left.

We have just completed another vibrotactile study that addresses some additional attentional issues. Subjects performed the task while looking at or away from the stimulated and responding hand that was placed in ipsilateral or contralateral hemispace (Bradshaw, Bradshaw, Pierson-Savage, & Nettleton, in press). In a simple RT version in which only one hand operated in one hemispace for a block of trials (thus permitting covert attention to be held to that side) there was a RSA, without effects from spatial compatibility or gaze direction; thus overt sustained gaze toward or away from a limb's locus of operation did not affect hemispatial asymmetries or general performance. We also tested subjects in a choice version of the task with hand/side uncertainty: either hand/side could be stimulated and have to respond, so that covert attention could not be directed successfully to either side alone, and the position of the responding limb had to be encoded despite the high stimulus–response compatibility of the task. The hands either lay in their own (ipsilateral) hemispace, or both crossed the midline to contralateral hemispace. With this choice design, a different pattern of results emerged: hemispace asymmetries disappeared, there were powerful hand-hemispace spatial compatibility effects, and performance was faster when subjects looked at (rather than away from) the stimulated or responding hand. However, this effect of gaze direction appeared in the crossed-arm condition only. Thus with response uncertainty, looking at the responding limb may resolve coding conflicts under conditions of spatial incompatibility.

Vibrotactile asymmetries may therefore stem from differences in our ability to hold covert attention to one or other side for a block of trials, and may reflect a difficulty among right-handed persons in focusing covert attention in far left hemispace, which is outside their usual functional processing space. Uncertainty about the spatial location of the stimulus and/or response, which leads to rapid shifts in such directed attention, may destroy hemispace asymmetries but may then generate hand–hemispace spatial compatibility effects. Overt sustained-gaze direction toward or away from the hemispace wherein a limb operates does not affect hemispatial asymmetries when the latter are present; when they are absent, under conditions of location uncertainty, looking at the responding limb may resolve conflicts of coding in the spatially incompatible crossed relationship of limb and hemispace. We need to distinguish between overt and covert, sustained and transient aspects of attention. Indeed, vibrotactile asymmetries have a strong attentional component, stemming from differences in our ability to hold covert attention to various spatial loci, and are not just a consequence of hemisphere–hemispace mapping relationships:

1. Performance in most of left hemispace is no worse than that in right hemispace; only the far left regions are disadvantaged.
2. It is speed of response initiation rather than stimulus detection that varies as a function of spatial position—and it is unlikely that the left hemisphere is specialized for something as simple as button pressing.
3. The hemispace account cannot explain why vibrotactile asymmetries disappear with low intensity stimuli. Likewise, it cannot easily explain the loss of hemispace effects with random-as-to-side presentation when attention is divided or must be shifted rapidly between the two sides of space.
4. Rehabilitated patients with hemineglect show left-side advantages only when their eyes are open, due presumably to the adoption of a visually based compensation strategy; performance reverts to a large RSA with eyes closed.

Of course the traditional anatomic connectivity model cannot explain our demonstrations that spatial location may determine performance asymmetries. Conversely, the fact that visual field and ear asymmetries readily occur to appropriate lateralized stimuli presented randomly to left and right sides (Bradshaw & Nettleton, 1983) indicates that such asymmetries cannot be the result of attentional factors alone. These asymmetries, like our auditory hemispace effects, could however be a consequence of hemisphere–hemispace relationships, rather than anatomic connectivity.

In conclusion it seems that the traditional anatomic-pathway-transmission model of lateral asymmetries needs a radical reappraisal. The hemispace effects (auditory, visual, tactual, and kinesthetic) that we have discussed here cannot be explained in terms of superior connectivity between one member of a pair of lateralized receptors–effectors and the hemisphere that is specialized for a particular task. Rather, the findings are perhaps better characterized as reflecting a functional relationship between each hemisphere and the contralateral half of space. It seems that it is the apparent spatial location of a stimulus (or the position of a responding limb) that may determine asymmetries, rather than just the proximal locus of receptor stimulation. This hemispace account should, however, be seen as complementing, rather than replacing the traditional anatomic model. Thus the brain may maintain two simultaneous maps: one which represents the distribution of stimuli on the proximal receptor surfaces, and the other map for relatively distal stimuli and responses that occur out beyond the body in extracorporeal space. Dissociation of these two maps may occur when the arms cross the body midline, when the head is turned relative to the body, or when the subject lies horizontally on one side. Although it is clear that the position of a stimulus may determine subsequent processing, the vibrotactile effects may not be the consequence of hemisphere–hemispace relationships. Rather, they may reflect attentional biases towards certain regions of space. The role of attention in these asymmetries needs further systematic investigation, but it now seems clear that any account of laterality effects needs to consider the role of anatomic connectivity, hemispace, and attention.

Acknowledgments. This work was supported by the Australian Research Grants Scheme (to J.L.B. and N.C.N.) and was written up while J.L.B. was on study leave at the Universities of St. Andrews and Exeter. We gratefully acknowledge the continued assistance in the design and construction of apparatus, of Bob Wood, John Dick, Geoff Mead, Bill Bramstedt and Noel Butson.

References

Anderson, R.A., Essick, G.K., & Siegel, P.M. (1985). Encoding of spatial location by posterior parietal neurons. *Science*, *230*, 456–458.

Autret, A., Auvert, L., Laffont, F., & Larmande, P. (1985). Electroencephalographic spectral power and lateralized motor activities. *Electroencephalography and Clinical Neurophysiology*, *60*, 228–230.

Baynes, K., Holtzman, J.D., & Volpe, B.T. (1986). Components of visual attention: Alterations in response pattern to visual stimuli following parietal lobe infarction. *Brain*, *109*, 99–114.

Bisiach, E., Capitani, E., & Porta, E. (1985). Two basic properties of space representation in the brain: Evidence from unilateral neglect. *Journal of Neurology, Neurosurgery and Psychiatry*, *48*, 141–144.

Bisiach, E., Comacchia, L., Sterzi, R., & Vallar, G. (1984). Disorders of perceived auditory lateralization after lesions of the right hemisphere. *Brain*, *107*, 37–52.

Bisiach, E., & Luzzatti, C. (1978). Unilateral neglect of representational space. *Cortex*, *14*, 129–133.

Blum B. (1985). Manipulation, reach and visual reach neurons in the inferior parietal lobule of the rhesus monkey. *Behavioral Brain Research*, *18*, 167–173.

Bonvillian, J.D., Orlansky, M.D., & Garland, J.B. (1982). Handedness patterns in deaf persons. *Brain and Cognition*, *1*, 141–157.

Boshoven, M.M., McNeil, M.R., & Harvey, L.O. (1982). Hemispheric specialization for the processing of linguistic and nonlinguistic stimuli in congenitally deaf and hearing adults: A review and contribution. *Audiology*, *21*, 509–530.

Bowers, D. & Heilman, K.M. (1980). Pseudoneglect: Effects of hemispace on a tactile line bisection task. *Neuropsychologia*, *18*, 491–498.

Bowers, D., Heilman, K.M., & Van Den Abell, T. (1981). Hemispace–VHF compatibility. *Neuropsychologia*, *19*, 757–765.

Bradshaw, J.L., Bradshaw, J.A., Nettleton, N.C., & Wilson, L.E. (1986). Leftwards error in bisecting the gap between two points: Stimulus quality and hand effects. *Neuropsychologia*, *6*, 849–855.

Bradshaw, J.L., Bradshaw, J.A., Pierson-Savage, J.M., & Nettleton, N.C. (in press). Overt and covert attention and vibrotactile reaction times: Gaze direction, spatial compatibility and hemispatial asymmetry. *Canadian Journal of Psychology*.

Bradshaw, J.L., Burden, V., & Nettleton, N.C. (1986). Dichotic and dichhaptic techniques. *Neuropsychologia*, *24*, 79–91.

Bradshaw, J.L., Nathan, G., Nettleton, N.C., Pierson, J.M., & Wilson, L.E. (1983). Head and body hemispace to left and right. III: Vibrotactile stimulation and sensory and motor components. *Perception*, *12*, 651–661.

Bradshaw, J.L., Nathan, G., Nettleton, N.C., Wilson, L.E., & Pierson, J.M. (1987). Why there is a left side underestimation in rod bisection. *Neuropsychologia*, *25*, 735–738.

Bradshaw, J.L., & Nettleton, N.C. (1981). The nature of hemispheric specialization in man. *Behavioral and Brain Sciences*, *4*, 51–63.

Bradshaw, J.L., & Nettleton, N.C. (1983). *Human Cerebral Asymmetry*. Englewood Cliffs, N.J.: Prentice Hall.

Bradshaw, J.L. Nettleton, N.C., Nathan, G., & Wilson, L.E. (1983). Head and body space to left and right, front and rear. II: Visuotactual and kinesthetic studies and left side underestimation. *Neuropsychologia*, *21*, 475–486.

Bradshaw, J.L., Nettleton, N.C., Nathan, G., & Wilson, L.E. (1985). Bisecting rods and lines: Effects of horizontal and vertical posture on left-side underestimation by normal subjects. *Neuropsychologia*, *23*, 421–425.

Bradshaw, J.L., Nettleton, N.C., Nathan, G., & Wilson, L.E. (1986). Tactual–kinesthetic matching of horizontal extents in the long term blind: Absence or reversal or normal left side underestimation. *Neuropsychologia*, *24*, 261–264.

Bradshaw, J.L., Nettleton, N.C., & Spehr, K. (1982). Braille reading and left and right hemispace. *Neuropsychologia*, *20*, 493–500.

Bradshaw, J.L., Nettleton, N.C., Wilson, L.E., & Bradshaw, C.S. (1987). Line bisection by left handed preschoolers: A phenomenon of symmetrical neglect. *Brain & Cognition*, *6*, 377–386.

Bradshaw, J.L. & Pierson, L.M. (1985). Vibrotactile reaction times in left and right hemispace: Stimulus and response uncertainty and gravitational and corporeal coordinates: In M.I. Posner & O.S.M. Marin (Eds.) *Attention and Performance XI*, (pp. 221–237). Hillsdale, N.J.: Erlbaum.

Bradshaw, J.L., Spataro, J., Harris, M., Nettleton, N.C., & Bradshaw, J.A. (in press). Crossing the midline by four to eight-year old children. *Neuropsychologia*.

Bradshaw, J.L., & Umilta, C. (1984). A reaction time paradigm can simultaneously index spatial compatibility and neural pathway effects: A reply to Levy. *Neuropsychologia*, *22*, 99–101.

Bryden, M.P. (1982). *Laterality: Functional Asymmetry in the Intact Brain*. New York: Academic Press.

Burden, V., Bradshaw, J.L., Nettleton, N.C., & Wilson, L. (1985). Hand and hemispace effects in tactual tasks involving interhemispheric integration in children. *Neuropsychologia*, *23*, 515–525.

Calford, M.B., Grayden, M.C., Huerta, M.E., Koas, J.H., & Pettigrew, J. (1985). A variant on the brain somatic map in a bat. *Nature*, *313*, 477–478.

Chedru, F. (1976). Space representation in unilateral spatial neglect. *Journal of Neurology, Neurosurgery and Psychiatry*, *39*, 1057–1061.

Cranney, T. & Ashton, R. (1982). Tactile spatial ability: Lateral performance of deaf and hearing age groups. *Journal of Experimental Child Psychology*, *34*, 123–134.

Crowne, D.P. (1983). The frontal eyefield and attention. *Psychological Bulletin*, *93*, 232–260.

De Renzi, E. (1982). *Disorders of Space Exploration and Cognition*. New York: Wiley.

De Renzi, E., Colombo, A., Faglioni, P., & Gilbertoni, M. (1982). Conjugate gaze paresis in stroke patients with unilateral damage. *Archives of Neurology*, *39*, 482–486.

Deutsch, D. (1980). The octave illusion and the what-where connection: In R.S. Nickerson (Ed.), *Attention and Performance VII* (pp. 575–592). New York: Academic Press.

Duhamel, J.R., Pinek, B., & Brouchon, M. (1986). Manual pointing to auditory targets: Performances of right versus left handed subjects. *Cortex*, *22*, 633–638.

Foley, J.M. (1975). Error in visually directed manual pointing. *Perception and Psychophysics*, *17*, 69–74.

Gardner, E. & Ward, A. (1985). Spatial compatibility in tactile–visual discrimination. *Neuropsychologia*, *17*, 421–424.

Geschwind, N. (1981). The perverseness of the right hemisphere. *Behavioral and Brain Sciences*, *4*, 106–107.

Gibson, C.J. & Bryden, M.P. (1984). Cerebral lateralization in deaf and hearing children. *Brain and Language*, *23*, 1–12.

Gilbert, C., & Bakan, P. (1973). Visual asymmetry in perception of faces. *Neuropsychologia, 11*, 355–362.

Halsband, U., Gruhn, S., & Ettlinger, G. (1985). Unilateral spatial neglect: A neuropsychological investigation. *International Journal of Neuroscience, 28*, 173–195.

Harris, L.J. (1980). Which hand is the eye of the blind? A new look at an old question: In J. Herron (Ed.), *Neuropsychology of Left Handedness*, (pp. 303–329). New York: Academic Press.

Harris, W.A. (1986). Learned topography: The eye instructs the ear. *Trends in the Neurosciences, 3*, 97–100.

Hartmann, W.M. (1983). Localization of sound in rooms. *Journal of the Acoustical Society of America, 74*, 1380–1389.

Heilman, K.M. (1979). Neglect and related disorders: In K.M. Heilman and E. Valenstein (Eds.), *Clinical Neuropsychology*, (pp. 268–307). Oxford: Oxford University Press.

Heilman, K.M., Bowers, D., & Watson, R.T. (1983). Performance on hemispatial pointing task by patients with neglect syndrome. *Neurology, 33*, 661–664.

Heilman, K.M., Bowers, D., & Watson, R.T. (1984). Pseudoneglect in a patient with partial callosal disconnection. *Brain, 107*, 519–532.

Heilman, K.M. & Valenstein, E. (1979). Mechanisms underlying hemispatial neglect. *Archives of Neurology, 5*, 166–170.

Heilman, K.M., Watson, R.T., & Valenstein, E. (1985). Neglect and related disorders. In K.M. Heilman and E. Valenstein (Eds.), *Clinical Neuropsychology*, (pp. 243–293). Oxford: Oxford University Press.

Heilman, K.M., Watson, R.T., Valenstein, E., & Damasio, A.R. (1983). Localization of lesions in neglect: In A. Kertesz (Ed.), *Localization in Neuropsychology*, (pp. 471–492). New York: Academic Press.

Hermelin, B. & O'Connor, N. (1971)(a). Functional asymmetry in the reading of Braille. *Neuropsychologia, 9*, 431–435.

Hermelin, B. & O'Connor, N. (1971)(b). Right and left handed reading of Braille. *Nature, 231*, 470.

Hewitt, W. (1962). The development of the human corpus callosum. *Journal of Anatomy, 96*, 355–358.

Jay, M.F. & Sparks, D.C. (1984). Auditory receptive fields in primate superior colliculus shift with changes in eye position. *Nature*, 309, 345–347.

Karavatos, A., Kaprinis, G., & Tzavaras, A. (1984). Hemispheric specialization for language in the congenitally blind: The influence of the braille system. *Neuropsychologia, 22*, 521–525.

Kimura, D. (1961). Cerebral dominance and the perception of verbal stimuli. *Canadian Journal of Psychology, 15*, 166–171.

Kimura, D. (1967). Functional asymmetry of the brain in dichotic listening. *Cortex, 3*, 163–178.

Kimura, D. (1976). The neural basis of language qua gesture: In H. Whitaker & H.A. Whitaker (Eds.), *Studies in Neurolinguistics, Vol. 1*, (pp. 145–156). New York: Academic Press.

Kimura, D. (1977). Acquisition of a motor skill after left hemisphere damage. *Brain, 100*, 527–542.

Kinsbourne, M. (1970). A model for the mechanism of unilateral neglect of space. *Transactions of the American Neurological Association, 95*, 143–146.

Kinsbourne, M. (1973). The control of attention by interaction between the cerebral hemispheres: In S. Kornblum (Ed.), *Attention and Performance IV*, (pp. 239–255). New York: Academic Press.

Kinsbourne, M. (1975). The mechanism of hemispheric control of the lateral gradient of attention: In P.M.A. Rabbitt & S. Dornic (Eds.), *Attention and Performance V*. (pp. 81–97). New York: Academic Press.

Larsen, S. & Hakonsen, K. (1983). Absence of ear asymmetry in blind children on a dichotic listening task compared to sighted controls. *Brain and Language*, *18*, 192–198.

Le Doux, J.E., Wilson, D.H., & Gazzaniga, M.S. (1977). Manipulospatial aspects of cerebral lateralization: Clues to the origin of lateralization. *Neuropsychologia*, *15*, 743–750.

Luh, K.E., Butter, C.M., & Buchtel, H.A. (1986). Impairments in orienting to visual stimuli in monkeys following unilateral lesions of the superior sulcal polysensory cortex. *Neuropsychologia*, *24*, 461–470.

MacKenzie, C.L. & Martenuik, R.G. (1985). Motor skill: Feedback, knowledge and structural issues. *Canadian Journal of Psychology*, *39*, 313–317.

Meredith, M.A. & Stein, B.E. (1985). Descending efferents from the superior colliculus relay integrated multisensory information. *Science*, *227*, 657–659.

Meredith, M.A. & Stein, B.E. (1986). Spatial factors determine the activity of multisensory neurons in cat superior colliculus. *Brain Research*, *365*, 350–354.

Mesulam, M.-M. (1981). A cortical network for directed attention and unilateral neglect. *Annals of Neurology*, *10*, 309–325.

Mesulam, M.-M. (1983). The functional anatomy and hemispheric specialization for directed attention. *Trends in the Neurosciences*, *6*, 384–387.

Mesulam, M.-M. (1985). Attention, confusional states, and neglect. In M.-M. Mesulam (Ed.), *Principles of Behavioral Neurology*, (pp. 125–168). Philadelphia: Davis.

Mishkin, M., Ungerleider, L., & Macko, K.A. (1983). Object vision and spatial vision: Two cortical pathways. *Trends in the Neurosciences*, *6*, 414–417.

Moscovitch, M. (1979). Information processing and the cerebral hemispheres: In M.S. Gazzaniga (Ed.), *Handbook of Behavioral Neurobiology, Vol. 2: Neuropsychology*, (pp. 379–446). New York: Plenum Press.

Nettleton, N.C. & Bradshaw, J.L. (1983). The effects of task, practice and sequencing upon the lateralization of semantic decisions. *International Journal of Neuroscience*, *20*, 265–282.

Neville, H.J., Kutas, M., & Schmidt, A. (1982). Event related potential studies of cerebral specialization during reading. II: Studies of congenitally deaf adults. *Brain and Language*, *16*, 316–337.

Ogden, J.A. (1985). Anterior–posterior interhemispheric differences in the loci of lesions producing visual hemineglect. *Brain and Cognition*, *4*, 59–75.

Ogden, J.A. (1987). The neglected left hemisphere and its contribution to visuospatial neglect. In M. Jeannerod (Ed.), *Neuropsychological and Physiological, Aspects of Spatial Neglect*. Amsterdam: North Holland.

Pandya, D.N. & Yeterian, E.H. (1984). Proposed neural circuitory for spatial memory in the primate brain. *Neuropsychologia*, *22*, 109–122.

Phillips, D.P. & Gates, G.R. (1982). Representation of the two ears in the auditory cortex: A re-examination. *International Journal of Neuroscience*, *16*, 41–46.

Pierson, J.M., Bradshaw, J.L., & Nettleton, N.C. (1983). Head and body space to left and right, front and rear I: Unidirectional competitive auditory stimulation. *Neuropsychologia*, *21*, 463–473.

Pierson-Savage, J.M. & Bradshaw, J.L. (in press). Mapping of extracorporeal space by vibrotactile reaction times: A far left side disadvantage. *Perception.*

Plourde, G. & Sperry, R.W. (1984). Left hemisphere involvement in left spatial neglect from right-sided lesions. *Brain*, *107*, 95–106.

Rakic, P. & Yakovlev, P.I. (1968). Development of the corpus callosum and cavum septi in man. *Journal of Comparative Neurology*, *132*, 45–72.

Ramos-Brieva, J.A., Olivan, J., Palomares, A., & Vela, A. (1984). Is there a right hemisphere dysfunction in major depression? *International Journal of Neuroscience*, *23*, 103–110.

Ratcliff, G. (1982). Disturbances of spatial orientation associated with cerebral lesions: In M. Potegal (Ed.), *Spatial Abilities: Development and Physiological Functions*, (pp. 301–333). New York: Academic Press.

Reinis, S., Landolt, T.P., Money, K.E., Lahue, R.H., & Weiss, D.S. (1986). Effect of head tilt on visual cortical cell function in the intact and labyrinthectomized cat. *Experimental Neurology*, *91*, 102–106.

Ritter, S.M. (1917). The vertical–horizontal illusion. *Psychological Review Monograph Supplement*, *23*(4), 1–99.

Rizzolatti, G., Gentilucci, M., & Matelli, M. (1985). Selective spatial attention: One center, one circuit or many circuits: In M.I. Posner & O.S.M. Marin (Eds.), *Attention and Performance XI*, (pp. 251–265). Hillsdale, N.J.: Erlbaum.

Rosenberger, P.B. (1974). Discriminative aspects of visual hemi-inattention. *Neurology*, *24*, 17–23.

Salamy, A. (1978). Commissural transmission: Maturational changes in humans. *Science*, *200*, 1409–1411.

Scarisbrick, D.J., Tweedy, J.R., & Kuslansky, G. Relationship between handedness and hand of performance on line bisection. (Unpublished manuscript).

Schenkenberg, T., Bradford, D.C., & Ajax, E.T. (1980). Line bisection and unilateral visual neglect in patients with neurologic impairment. *Neurology*, *30*, 509–517.

Schwartz, A.S., Marchok, P.L., Kreinick, C., & Flynn, R.E. (1979). The asymmetrical lateralization of tactile extinction in patients with unilateral cerebral dysfunction. *Brain*, *102*, 669–684.

Smith, Y.M. & Latto, R. (1982). An investigation of visual neglect following parietal lobe lesions in man. *Behavior and Brain Research*, *5*, 120–121.

Stein, J.F. (1986). Role of the cerebellum in the visual guidance of movement. *Nature*, *323*, 217–221.

Stuss, D.T. & Benson, D.F. (1984). Neuropsychological studies of the frontal lobes. *Psychological Bulletin*, *95*, 3–28.

Tressoldi, P.E. (1987). Visual hemispace differences reflect hemisphere asymmetries. *Neuropsychologia*, *25*, 625–636.

Walker, J.T. (1972). Natural visual capture in bilateral length comparisons. *Perception and Psychophysics*, *11*, 247–251.

Weiner, M.J., Hallet, M., & Funkenstein, H.H. (1983). Adaptation to lateral displacement of vision in patients with lesions of the central nervous system. *Neurology*, *33*, 766–772.

Weston, P. & Weinman, J. (1983). The effects of auditory and linguistic deprivation on lateral preferences of deaf children. *Developmental Medicine and Child Neurology*, *25*, 207–213.

Witelson, S.F. (1985). The brain connection. The corpus callosum is larger in left handers. *Science*, *229*, 665–667.

Yakovlev, P.I. & Lecours, A.R. (1967). The myelogenetic cycles of regional maturation in the brain: In A. Minkowski (Ed.), *Regional Development of the Brain in Early Life*, (pp. 355–358). Oxford: Blackwell.

2
Lateralization of Lexical Processes in the Normal Brain: A Review of Visual Half-field Research

CHRISTINE CHIARELLO

The extent to which the right hemisphere can subserve language functions is one of the most hotly debated issues in neuropsychology today, as witnessed by the lively interchanges which have appeared in several recent journals (Gazzaniga, 1983a, 1983b; Levy, 1983; Zaidel, 1983b; Myers, 1984; Patterson & Besner, 1984a, 1984b; Rabinowicz & Moscovitch, 1984; Zaidel & Schweiger, 1984). A focal issue in this controversy concerns right hemisphere involvement in the recognition of visually presented words, and much of the debate has turned on differing interpretations of evidence from commissurotomy and acquired dyslexia. Some have argued for extensive participation of the right hemisphere (Coltheart, 1980a; Zaidel, 1983b), while others find little reason to presume any significant role for the right hemisphere (Patterson & Besner, 1984a, 1984b; Gazzaniga, 1983a, 1983b). In the context of such discussions it is somewhat surprising that evidence from lateralization studies with neurologically normal individuals has been given only piecemeal consideration. For more than a decade researchers have been exploring the consequences of selectively stimulating the cerebral hemispheres by using tachistoscopic methods, with words being the most frequently employed stimuli. Any comprehensive model of hemisphere differences in lexical processing must encompass the results of such studies in addition to the evidence from neurologically compromised subjects.

To the extent that studies of normal subjects have been considered, the following approach has been taken. In light of some rather dramatic dissociations observed among patients with deep dyslexia and commissurotomy (i.e., between reading of abstract and concrete words, and the presence or absence of phonological transcoding skills), investigators looked to the literature on lateralization in normal subjects to determine if such functions might be laterally dissociable in the normal brain. Despite some early encouraging findings (Ellis & Shepherd, 1974; Cohen & Freeman, 1978), subsequent studies at best yielded mixed results and at worst were nearly impossible to interpret. From this it was concluded that no coherent pattern of results emerged from visual half-field (VHF) studies (Patterson & Besner, 1984a), that there is no typical profile of right hemisphere (RH) reading demonstrable in neurologically normal individuals (Rabinowicz & Moscovitch, 1984), and thus that there is little hope that visual laterality research can meaningfully contribute to the "right hemisphere debate."

On the surface this pessimistic conclusion seems unassailable. However, the relation between dissociations observed in abnormal reading states and the normal reading system may not be as clear-cut as one might like. We should at least allow for the possibility that some of the variability in the laterality findings may be attributable to an attempt to pin down phenomena which are irrelevant to normal word recognition (in either hemisphere), rather than to intrinsic problems with lateralization techniques. Though we all hope to find parallels across populations, a lack of convergence does not necessarily invalidate either body of data. It may also indicate that somewhat different mechanisms underlie the behavior of normal and brain-injured readers.

A second reason to question the above conclusion relates to the selective manner in which visual laterality findings have been brought to bear on these issues. Although there are many studies that examine the recognition of abstract versus concrete words, lateralization researchers have actively pursued other phenomena that are no less germane to the question of hemispheric specialization in reading. The former results are widely cited, but many of the latter findings are less well-known. If the goal is to determine whether words are processed in characteristically different ways in the normal left and right hemispheres, then this is best determined by a systematic consideration of all available visual lateralization research. The relation of such "hemispheric profiles" (if demonstrable) to findings from pathology is a separate (and logically subsequent) question.

This chapter follows an information-processing approach to review the visual half-field (VHF) research relevant to the issue of lexical lateralization. This approach is based on two assumptions. The first is that word recognition is a complex process, involving the interaction of numerous subprocesses and components that can be selectively studied with information-processing techniques. The second is that the hemisphere asymmetry observed in a given situation is multiply determined. For example, when a right visual field (RVF) advantage for the pronunciation of a printed word occurs, this is simply the culmination of a number of operations that in sum are more efficiently performed when processing is initiated by the left hemisphere (LH). What remains to be discovered are the precise hemispheric substrates for each of these operations. A RVF advantage may arise from an asymmetry for a relatively early process (with subsequent operations being equally available to either hemisphere). It might also reflect hemispheric equivalence for processes that precede and follow lexical access, coupled with predominance of the left hemisphere in the process by which lexical entries are retrieved. Additional possibilities are equally consistent with this result.

The only way to disentangle these factors is by comparing performance across tasks, or within a task across stimuli–processing conditions whose information processing requirements are fairly well understood. In so doing one can draw on an impressive range of techniques and findings that have resulted from decades of research on the mechanisms of normal word recognition. These findings provide the framework for interpreting the VHF studies that are reviewed below.

It is useful to begin by discussing the information-processing requirements of the most frequently used VHF lexical tasks. Next a "bare bones" processing model is presented, which will make explicit the major subprocesses involved in these tasks. Then the VHF results are examined, beginning with those most relevant to the earliest processes (i.e., visual analysis and encoding), and continuing with consideration of effects believed to have a later locus. For each effect examined we ask whether it is present to the same extent in both visual fields or is quantitatively or qualitatively different depending on the hemisphere of input. Thus it is possible to build up performance profiles for each hemisphere based on the entire sequence of information processing mechanisms believed to underlie visual word recognition in normal readers. Only then can be results be validly compared to other neuropsychological populations. For this reason the clinical neuropsychological evidence is not considered in deriving the reading profile of each hemisphere.

This review attempts to include all published VHF studies meeting the following criteria:

1. Use of word or word-like (i.e., pronounceable letter strings) stimuli. Investigations involving single letters, picture naming, word–picture interference, and other similar phenomena were not included as their relevance to the issue of word recognition in reading is unclear. For the same reason, studies involving the processing of words held in memory are not considered.
2. Subjects who were neurologically normal, right-handed adults. Excluded were studies of children, bilinguals, or disabled readers. Those comparing across subject groups (e.g., left- and right-handed persons) were included provided they met the other criteria. However, only data for the right-handed individuals were examined.
3. Consideration of at least one information-processing variable (e.g., exposure time, word frequency, lexical priming). Studies that did not manipulate processing variables (e.g., those using one word recognition measure as a laterality index) provide little basis for inferring the mechanisms underlying word recognition in each hemisphere, and thus were not included.

Lexical Processing Mechanisms in Simple Reading Tasks

Visual half-field investigations of lexical processing rely almost exclusively on three tasks: naming, lexical decision, and word judgments. In naming the subject is required to pronounce a visually presented word; in lexical decision to discriminate words from orthographically regular, pronounceable letter strings ("legal" nonwords); in word judgment, the subject determines whether a word or pair of words possesses a particular property (e.g., is it a type of furniture? do the words rhyme?). Although superficially quite different, these tasks have in common many lexical processes and differ in specifiable ways.

It is useful to differentiate three major classes of information processing operations required to perform these tasks. Although there is no universal agreement on the exact boundaries between these classes, some such division is implicit in most current models of lexical processing (see Seidenberg, 1985). *Prelexical* processes are those involved in visual–sensory analysis of the stimulus and subsequent encoding operations. The outcome of such processes is an abstract representation that is used to access lexical information in memory. Variables that alter stimulus quality (such as exposure time) or affect ease of encoding (such as word length) will predominantly affect prelexical operations. If the hemispheres differ in prelexical processes, we should find differential effects of such variables in each visual field.

The *lexicon* is the mental storehouse for all information about individual word, such as their meanings, pronunciation, and syntactic features. *Lexical processes* are those that mediate access to this store and culminate in the retrieval of lexical information and comprehension. Some variables, such as word frequency, are thought to have their primary effect on the lexical access process. One could then investigate possible hemisphere differences in lexical access by presenting words of different frequencies and observing whether changes in asymmetry result.

Postlexical processes are those subsequent to lexical retrieval in which the retrieved information is acted on in some way: for example the pronunciation process in naming, or the decision stage in lexical decision. Because lexical access and comprehension result in no discernable behavior, all tasks used to assess word recognition will also involve some postlexical processing—there is no "pure" measure of word comprehension[1]. Variables that affect postlexical processing will differ according to the task. In lexical decision, for example, some types of semantic priming have been shown to have predominantly postlexical effects (Seidenberg, Waters, Sanders, & Langer, 1984).

When comparing these tasks, it is evident that they should involve similar prelexical and lexical processes (i.e., they all require lexical access), but the tasks should differ considerably in their postlexical processing requirements. In naming, postlexical processes involve, at a minimum, conversion of the retrieved pronunciation code into a sequence of articulatory acts, and may also include a contribution from nonlexical pronunciation codes

---

[1] However, evoked potentials may ultimately provide a means of assessing comprehension in the absence of any overt response.

(see Rosson, 1985 and discussion below). In lexical decision, postlexical stages include weighing the evidence that the letter string is a word (i.e., whether a "matching" lexical representation has been accessed, or no match has been found) according to some decision criterion (Gordon, 1983), and signaling the decision once made. Because word-judgment tasks typically involve comparing two words along some dimension (usually semantic), these tasks demand considerable postlexical processing: for example, a comparison of the retrieved semantic information and related decision processes. Because of their relative complexity, word-judgment tasks are rarely used to assess the earlier stages of word recognition, but may indicate the extent to which the hemispheres differ in postlexical operations.

Lexical decision and naming can be used to infer prelexical and lexical access mechanisms, provided we are cautious in our interpretations. If a variable (such as word length) has its effect on processes prior to lexical retrieval, it should have similar effects in lexical decision and naming. Conversely, a variable with differential effects on these tasks is likely to have a postlexical locus (because lexical decision and naming differ primarily in the processes they recruit subsequent to lexical access). These points will be made more explicit by considering a simplified processing model (see Figure 2.1).

Although virtually all models of the lexicon are considerably richer than that depicted in Figure 2.1, a more primitive framework is sufficient for examining the results of VHF studies. There are several aspects of the model that require comment. The first concerns the role of phonological coding, a topic that has engendered considerable debate among investigatiors of reading (Coltheart, 1978; McCusker, Hillinger, & Bias, 1982). Three "routes" for the conversion of print to sound have been proposed: a *prelexical* route in which graphemic representations are phonologically encoded prior to lexical access; a *lexical* route in which phonology is retrieved (or constructed, see Glushko, 1979) from information stored in the lexicon; and a *nonlexical* route in which phonology is derived via grapheme–phoneme correspondences and does not involve lexical accesss. The prelexical route is not represented in the model because there is little evidence that phonological encoding occurs prior to lexical access in standard reading tasks (for discussed see Coltheart, 1980b; Banks, Oka, & Shugarman, 1981). The relative contributions of lexical and nonlexical information to the pronunciation of words and nonword strings are contested: some argue that even nonwords could be pronounced via lexical analogies (Glushko, 1979), and others opt for a nonlexical process for nonwords and the lexical route for words (Coltheart, 1980b). Because the most recent evidence suggests that both routes are variably involved in the pronunciation of both words and nonwords (Koriat, 1984; Rosson, 1985), the model postulates a postlexical pronunciation mechanism that allows these two sources of information to interact in deriving the final arti-

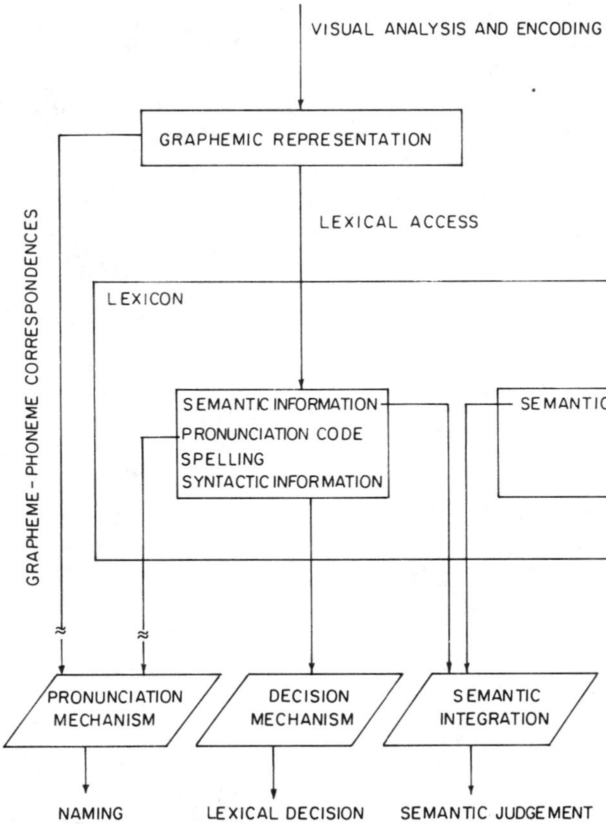

FIGURE 2.1. A simplified model of the major subprocesses involved in VHF lexical tasks.

culatory code. It should be noted that some postlexical mechanism is necessary even if all pronunciation is lexically based (Glushko, 1979): information from activated lexical analogies must somehow be pooled in constructing the pronunciation (Henderson, 1982).

A second aspect of the model concerns the structure of the lexical representations themselves. Figure 2.1 depicts the scenario in which access to a single lexical entry makes available all stored information about the word. However, some models posit separate storage of phonological, orthographic, and semantic information with access to one store directing access to corresponding information in the other stores (Forster, 1978; Morton & Patterson, 1980). As the studies reviewed here are neutral with respect to this distinction, the former was selected for illustrative purposes only. The model also makes explicit that lexical decisions are based on the presence or absence of an entry in the lexicon, but that retrieval of semantic or other lexical information need not occur. In contrast, lexical judgments such as decisions about semantic category membership require semantic retrieval; Figure 2.1 depicts a situation in which a judgment is required about whether two words are semantically similar.

Using this model as a framework we can begin to consider whether the VHF results indicate differential hemispheric mediation of various lexical subprocesses.

Prelexical Processes

Stimulus Quality Variables

In the earliest stages of tachistoscopic word recognition the features needed for further stimulus encoding and recognition must be extracted from a briefly presented, laterally displaced visual display. Although this places high demands on early visual processes, until quite recently it was assumed that both hemispheres were equally adept at such operations, and that any observed asymmetry could be attributed to subsequent processes. Sergent (1983a, 1983b) has disputed this and presents evidence that at least some task asymmetries can be due to differential hemispheric sensitivity to the quality of the sensory input. If VHF lexical asymmetries are partly determined by hemisphere differences in such purely visual processes, we should be able to alter the asymmetry by varying stimulus quality.

There are few investigations of visual quality variables with lexical stimuli. When stimulus exposure time is varied (range, 6–150 ms) the RVF advantage for word recognition remains unaltered in both lexical decision (Regard, Landis, & Graves, 1985; Chiarello, Senehi, & Soulier, 1986) and naming (Gill & McKeever, 1974; Mackavey, Curcio, & Rosen, 1975; Young & Bion, 1980; Young, Bion, & Ellis, 1980).[2] In these studies the direction and magnitude of the RVF advantage remained constant, despite substantial changes in overall performance levels induced by varying exposure time. The sole opposing finding is that of Pring (1981) who reported a LVF advantage with exposures of 50 ms, but a RVF advantage at 150 ms for lexical decision latencies. However, the error data were contradictory with a RVF advantage at 50 ms, but no asymmetry at 150 ms. Thus the apparent exposure-dependent change in asymmetry can be attributed to speed–accuracy trading,[3] and Pring's results can only be regarded as inconclusive. In general, it appears that the hemispheres are equally sen-

---

[2] The results of Bradshaw, Hicks, and Rose (1979) are not included because the nonword strings in their "lexical" decision experiment were illegal (random letter permutations), and thus the task did not require lexical processing

[3] Such speed–accuracy tradeoffs are rare in the studies reviewed here; comparable results are nearly always found regardless of whether reaction time or accuracy is measured. For the sake of economy I only distinguish between such measures when comparable results are *not* found.

sitive to visual quality to the extent that this depends on exposure duration. Similar findings have been reported for variations in retinal eccentricity: the RVF advantage for lexical decision remains constant despite the diminished stimulus quality which occurs with increases in eccentricity (Babkoff, Genser, & Hegge, 1985; Chiarello et al., 1986).

Sergent (1984), in a study using the naming task, manipulated visual quality by comparing high- and low-contrast stimuli. The RVF advantage for high-contrast words was significantly smaller than that found for low-contrast words (lower contrast being more disruptive to LVF than RVF stimuli). It is difficult to know whether this reflects a hemispheric difference in the ability to extract information from a degraded signal or whether, as Sergent argues, stimulus intensity information is preserved when LVF input is transferred to the left hemisphere for the vocal response. The former interpretation would be favored if similar results were found with a nonvocal lexical task; at present either interpretation is tenable. However, there is one procedural variable in the Sergent (1984) study that adds a further complication. Her subjects participated in three 160-trial sessions, yet only four different words were used in the study. This means that subjects viewed each of the four stimuli 120 times. It would be difficult to argue that the visual identification processes used in this situation (where subjects could rely on a small set of critical features) are typical of those normally employed in recognizing words. Thus the significance of the enhanced asymmetry for low-contrast words is uncertain.

Visual processing can also be disrupted by following the stimulus with a mask. Unstructured (flash or noise) masks are thought to disrupt relatively early sensory processes, while patterned masks interfere with later stages in which sensory features are available for perceptual encoding (Turvey, 1973). The time course over which masking is effective is determined by varying the interval between presentation of the stimulus and mask. The length of the masking interval provides one measure of the time necessary for sensory (unstructured mask) versus perceptual (pattern mask) features to be extracted. Thus, comparisons between mask types and their time courses allow us to infer whether sensory or later perceptual encoding stages proceed in a similar fashion in each hemisphere. Though unstructured masks have equivalent effects in each visual field (Moscovitch, 1983), patterned masks disrupt processing over greater stimulus–mask intervals in the LVF than in the RVF (Oscar-Berman, Goodglass, & Cherlow, 1973; Michaels & Turvey, 1979, experiment E2; Moscovitch, 1983). This implies that the extraction of critical *perceptual* features for words requires more time for LVF than for RVF stimuli. Thus, when the same target–mask interval is used in each visual field, pattern masking may accentuate the RVF advantage because LVF performance suffers greater degradation under these conditions (Moscovitch, 1983; Lambert & Beaumont, 1983, experiment 2; Byrd & Moscovitch, 1984; Sergent, 1984; Young & Ellis, 1985, experiments 1, 2).

These findings suggest that initial visual–sensory analyses are comparable in each hemisphere, but that the process of encoding visual information into the orthographic features necessary for word recognition requires more time, and is more easily disrupted, in the LVF–RH. Theoretically this LVF degradation could be attributed to the process of interhemispheric transfer (Moscovitch, 1983). However, other research (reviewed below) suggests that word encoding proceeds in a qualitatively different manner in each visual field, which casts doubt on the callosal transfer view.

Given the relative dearth of research on stimulus quality in lexical tasks, only provisional conclusions are possible. At present, it appears that the hemispheres are equally adept at visual–sensory operations (sensory quality variables do not differentially effect the visual fields), but that a RVF advantage begins to emerge when visual features are available for encoding into more abstract perceptual representations (the process disrupted by pattern masking).

Orthographic Structure Variables

Variables such as word length, orientation, and script or typeface can be expected to affect the ease with which words are encoded.[4] In this section we consider the role of such factors which are intrinsic to the recognition of words (as opposed to other visual stimuli), to infer whether the hemispheres differ in prelexical encoding operations.

Word Length and Word Superiority Effects

In naming tasks, performance in the LVF (but not in the RVF) systematically varies with the length of the word (Gill & McKeever, 1974; Schiepers, 1980; Bub, 1982, experiment 1; Young & Ellis, 1985). Thus the magnitude (and in some cases the presence) of the RVF advantage is dependent on word length. In lexical decision studies examining string length, similar results are found: LVF performance declined with longer words, but RVF performance did not change (Leiber, 1976; Bûb, 1982, experiment 3).

In an impressive series of experiments, Young and Ellis (1985) determined the boundary conditions for this effect and ruled out a number of potential artifactual causes. Because stimuli in the above studies were horizontally presented, length effects in the LVF might be attributable to a fall-off in acuity for initial letters, which would be accentuated as word length increased. However, when short and long words were presented such that their initial letters appeared in the same retinal locations, the length *X* visual field interaction was unaltered (Young & Ellis, 1985, ex-

---

[4] Some very recent evidence on word recognition in Dutch (Hudson & Bergman, 1985) implies that length effects in lexical decision may be attributable to a post-access spelling check. However, the similarity in visual field length effects for lexical decision and naming, and their restriction to LVF stimuli (as reviewed below), argue against a postlexical locus for the length effects described here.

periment 3). These investigators further demonstrated that the interaction was dependent on number of letters rather than syllabic length (experiment 4), and held equally for words of different imageability or frequency classes (experiments 5, 6), or when overall performance levels were diminished (experiment 2, thus ruling out a RVF ceiling as the source of the interaction).

These findings imply that letter strings are processed serially, one letter at a time in the LVF, but in parallel in the RVF. However a further experiment suggests an important modification to this position. When subjects were asked to pronounce nonsense syllables, there were equivalent length effects across the visual fields: performance was better in the RVF overall, but accuracy in both VFs declined with string length (Young & Ellis, 1985, experiment 7). Thus it appears that any letter string presented to the LVF is encoded in a serial, letter-by-letter manner, while words in the RVF are encoded as a unit. The latter finding may indicate that left hemisphere lexical access occurs before letter identification has been completed. Nonsense strings, having no lexical representations, are processed in a comparable letter-by-letter fashion whatever the hemisphere of input.

There was only one condition under which Young & Ellis could demonstrate word length effects in the RVF. When words were presented vertically, or when the letters were misaligned, performance in both the RVF and the LVF declined with increasing length (experiment 8). Bûb (1982, experiment 2) also obtained comparable length effects across the visual fields for vertical, but not horizontal, words. These findings are very important because they indicate that specialized word encoding mechanisms in the RVF left hemisphere may only be invoked when words are presented in a standard (horizontal) format; when words are displayed in a visually unfamiliar fashion the *qualitative* visual field difference in word encoding disappears. The implications of this finding for the interpretation of visual field effects with vertically arrayed words is discussed further below.

There are several conclusions to be drawn from these experiments. First, under standard presentation conditions, words are encoded in a qualitatively different manner depending on the field of input, and only in the RVF are words "special" (i.e., being processed differently from nonword letter strings). This implies that words are primarily encoded in the hemisphere of input. As Young and Ellis (1985) argue, were LVF information transferred to the left hemisphere prior to such encoding, only quantitative visual field differences should have been observed (thus, no length X visual field interaction). Second, visual field asymmetries can be manipulated simply by altering word length. In two of the four experiments in which Young and Ellis (1985) employed three-letter horizontal words, no visual field difference was obtained for these stimuli. Presumably, letter-by-letter (LVF) encoding of three-letter words is not invariably less efficient than whole-word (RVF) encoding. (Absent or reversed asymmetries in lexical tasks do seem to be more common when three-letter words are used; see Bradshaw, Hicks, & Rose, 1979; Young, Bion, & Ellis, 1980, experiment 3; Moscovitch, 1983.) This further suggests that a large portion of the RVF advantage for words may be attributable to hemispheric differences in encoding procedures.

Word superiority effects provide converging evidence that only RVF stimuli benefit from specialized word-encoding processes. Such effects are demonstrated by superior recognition of a letter when embedded in a word, than when the letter occurs in isolation or within a nonsense string (see Baron, 1978; Henderson, 1982 for reviews). Typical word superiority effects are present for stimuli presented to the RVF, but are absent for letter strings input to the LVF (Kreuger, 1975; Besner, 1983). Thus letter identification in the LVF is not enhanced when a word context is available. Because word superiority effects generally survive visual mutilations such as case alternation (Purcell & Stanovich, 1982), it is not surprising that the RVF word superiority effect occurred for both vertically and horizontally presented strings (Kreuger, 1975).

These findings suggest that word encoding in the RVF is not mediated by prior identification of component letters, and that a lexical code is available to assist in the processing of individual letters. However, there is no evidence that lexical codes are available prior to letter identification in the LVF. This is consistent with the view that LVF word encoding is mediated by a serial, letter-by-letter process. Lexical access in the LVF may depend on previous identification of the word's component letters. In any case, lexical information does not facilitate letter recognition in the LVF as it does in the RVF.

Further support for a serial word encoding process in the LVF comes from a study in which visual identity judgments were made to two simultaneously presented words (Brand, Van Bekkum, Stumpel, & Kroeze, 1983). On trials where the words did not match, LVF performance was found to depend on the position of the differing letter: accuracy decreased and reaction time increased as the differing letter changed from initial to medial to final word position. However, no position effects were found for RVF stimuli. Once again the evidence suggests that words are encoded and examined as whole units in the RVF–LH, but as serial letter strings in the LVF–RH.

Orientation and Script

The findings of Young and Ellis (1985) imply that the use of vertically oriented stimuli disengages top–down whole-word encoding procedures normally employed in the RVF. Because, with vertical presentation, length effects are found for both RVF and LVF words, it appears that a serial encoding process is invoked in both hemispheres by virtue of the unusual presentation format. Thus we might expect vertical word presentation to disrupt RVF more than LVF processing.

Vertical orientation has often been employed to "control" for postexposural left-to-right scanning that, if used, would favor horizontal letter strings in the RVF. It is clear that the RVF advantage does not depend on the use of horizontal words (see Barton, Goodglass, & Shai, 1965; Saffran, Bogyo, Schwartz, & Marin, 1980; Boles, 1985), and there is little direct evidence that scanning contributes to asymmetrical processing of horizontally presented words (see Bradshaw, Nettleton, & Taylor, 1981b; Tomlinson-Keasey, Brewer, & Huffman, 1983; Chiarello, in press for discussion). However, if vertical presentation forces the RVF–LH to adopt a less efficient encoding process (one not qualitatively different from that in the LVF–RH), we might expect to see less pronounced or less reliable asymmetries with vertical words.

Some investigators report comparable visual field asymmetries with horizontally and vertically arrayed words (Mackavey, et al., 1975; Ellis & Young, 1977; Boles, 1985; Young & Ellis, 1985, experiments 1, 8). Others find a much smaller RVF advantage with vertical words (McKeever, 1971; McKeever & Gill, 1972; Lambert & Beaumont, 1983, experiment 1). In the latter studies, the diminished asymmetry was primarily attributable to a pronounced drop to RVF performance with vertical presentation. All of these experiments used the naming task. No direct horizontal–vertical comparisons have been made with lexical decision, but it is notable that visual field asymmetries were absent in the lexical decision studies of Shanon (1979a, 1979b) and Day (1977, 1979—for concrete words) where vertical presentation was employed. In contrast, comparable studies with horizontal strings report reliable RVF advantages for words (see Leiber, 1976; Chiarello, Dronkers, & Hardyck, 1984; Hardyck, Chiarello, Dronkers, & Simpson, 1985a).[5]

Thus, there is some support for the view that the use of vertically oriented words may diminish (in naming) or abolish (in lexical decision) the typical RVF advantage. However, this does not imply that the RVF advantage seen with horizontal words is due to postexposural scanning. If we take the existence of word length effects as a behavioral index of a serial, letter-by-letter "scanning" process, then the following conclusions emerge:

[5] I am only aware of one lexical decision study using horizontal strings in which no visual field asymmetries were found (Chiarello, Church, & Hoyer, 1985). In this experiment, where stimuli were shown on a microcomputer screen, no laterality effects were found, despite the extremely robust RVF advantages foud in all our previous and subsequent lexical decision studies (a total of more than 20 separate experiments). In all our other studies stimuli were presented as high-contrast slides or using a vector graphics display system, yet there is no reason to attribute the negative findings to the apparatus. This remains a puzzling and embarrassing result, which we are still unable to explain. The possibility of a type I error cannot be discounted.

(*a*) All letter strings input to the LVF–RH, regardless of orientation, will be encoded via a serial scan. (*b*) *Words* input to the RVF–LH can be encoded as a unit, without letter-by-letter processing, if they are presented in a standard, horizontal format. However, with vertical or other unfamiliar presentation, specialized LH lexical encoding mechanisms are not invoked, and the word will be processed serially, in a manner not qualitatively different from that used by the right hemisphere. (*c*) Smaller or more labile RVF advantages with vertical presentation reflect the inability of the LH to invoke its specialized word encoding processes. Thus the use of vertical words actually increases the likelihood of serial scanning, and orientation should not be treated as a hemispherically "neutral" variable in lexical VHF studies.

A final variable that may effect word encoding processes is the script or typeface in which the letter strings appear. Given the well-known VHF dissociations between the processing of Japanese Kana and Kanji scripts (Hatta, 1977a, 1977b; Sasanuma, Itoh, Mori, & Kobayashi, 1977), and the influence of typeface on asymmetries for letter recognition (Bryden & Allard, 1976), it is rather surprising that the role of such variables in the processing of lexical alphabetic stimuli is virtually unexplored.

Sergent (1984) compared asymmetries for upper- and lower-case words in a naming task. The RVF advantage was larger for upper-case than for lower-case words, but only under masking conditions. Sergent interpreted this to indicate that, when stimulus quality is poor, the right hemisphere is better able than the left to make use of configurational cues (present with lower-case stimuli) to aid in word recognition. However, we must reiterate the previously mentioned qualifier that numerous repetitions of the same four stimulus words were used in this study. It is unlikely that configurational cues would be of much value with a larger stimulus set, and thus the ultimate generalizability of this finding is questionable. Indeed, Boles (1985, experiment 1) found a RVF advantage of equal magnitude for naming upper- versus lower-case pattern-masked words (each stimulus reappearing four times throughout the experiment).

In a lexical decision task, Regard et al. (1985) compared words printed in lower case letters to the same stimuli written in more cursive stenography (only skilled stenographers served as subjects). Printed words yielded the typical RVF advantage over a range of exposure times. For stenographic words, however, a RVF advantage was obtained with 150-msec exposures, but a LVF advantage with 50-msec exposures. This would suggest that there is some aspect of stenographic script that is more readily encoded by the right hemisphere when processing time is especially brief.

No firm conclusions can be drawn on the basis of only three studies, but there is some suggestion that the RH may play a greater role in the processing of some typefaces or scripts under rather severe data limitations (such as masking or brief exposures). This implies that the features the RH is able to extract from very degraded stimuli may be more informative (i.e., lead to more accurate identification) for some scripts, than the features extracted by the LH. If future investigations confirm these findings, we may have to modify the interpretation offerred earlier (of pattern-masking results) that critical perceptual features for words are extracted more efficiently in the RVF. The asymmetry for this process may to some extent depend on the nature of the script.

Summary

Review of the data relevant to prelexical word recognition processes indicates that early visual–sensory operations proceed in a similar manner in each hemisphere. Hemisphere differences begin to emerge when visual features become available for perceptual encoding. In most circumstances it appears that the process of extracting critical perceptual features is performed more efficiently in the left hemisphere. There is converging evidence that fundamentally different encoding processes operate on words input to the right and left hemispheres. We cannot escape the conclusion that a serial, letter-by-letter process is used to encode words in the LVF. This process becomes more and more inefficient as word length increases. Words in the LVF are processed no differently than other letter strings, and thus a lexical context does not facilitate recognition of component letters. In contrast, words have a priviledged status when directly input to the left hemisphere. As long as they are presented in a standard reading format they are immediately encoded as a perceptual unit. Furthermore, word identity constrains identification of component letters rather than vice versa. It is evident that word encoding in the left hemisphere is aided by lexically specialized processors, while the right hemisphere uses more general purpose algorithms. Although the latter are generally less efficient, they may prove to be advantageous when words are presented in very unusual or visually complex formats (perhaps due to "overspecialization" of LH processors). The next question to be addressed is how these differentially encoded representations are used to access the lexicon.

Lexical Processes

Word Frequency and Related Effects

We can now consider whether the hemispheres differ in the manner in which lexical entries are accessed. The *sine qua non* of a variable affecting this process is word frequency: high-frequency words are recognized more quickly and accurately than are those of lower frequency in both lexical decision and naming tasks. This effect is taken to reflect either a frequency-ordered lexical search (Stanners & Forbach, 1973; Becker, 1976), or frequency-sensitive detection thresholds for lexical representations (logogens) (Morton, 1969; Coltheart et al., 1977). Though the most recent evidence suggests that word frequency may also effect postlexical processes in both lexical decision (Balota & Chumbley, 1984) and naming (Balota & Chumbley, 1985), it is likely that this variable has its primary locus at the lexical access stage.

It is often asserted that the right hemisphere can only recognize familiar, high-frequency words, and that hemispheric differences will be greatest when low-frequency words are presented (e.g., Bradshaw, 1980; Goodall, 1984). Visual half-field studies with normal subjects provide no empirical support for this claim. Substantial and equivalent frequency effects are found in both visual fields. Consequently, when high- and low-frequency words are compared, there is no corresponding change in the direction or magnitude of the RVF advantage, either in lexical decision (Leiber, 1976; Bradshaw & Gates, 1978, experiment 1; Bradshaw, Nettleton, & Taylor, 1981a) or in naming (Caplan, Holmes, & Marshall, 1974; Hines, 1976, 1977; Orenstein & Meighan, 1976; Bradshaw & Gates, 1978, experiment 3; Bradshaw & Taylor, 1979; Boles, 1983; Young & Ellis, 1985, experiment 6). Likewise the LVF advantage for the recognition of Kanji is similar for high- and low-frequency words (Hatta, 1977b). The equivalence in frequency effects across the visual fields persists even when words are chosen from the extremes of the frequency distribution (Bradshaw & Gates, 1978, experiment 1; Bradshaw & Taylor, 1979), or when high-frequency concrete words are compared with low-frequency abstract words (Bradshaw et al., 1981a). Because robust visual field and frequency main effects were found in all of the above studies (only Orenstein & Meighan [1976] did not find a significant word frequency effect), it is unlikely that the failure to find a visual field–frequency interaction is due to lack of statistical power. These results suggest that frequency-dependent aspects of lexical access are comparable regardless of the hemisphere of input.

Two apparent exceptions to this conclusion are the studies of Tomlinson-Keasey, et al. (1983) and Heister (1983; Heister, Kolbert, & Hofmeister, 1983). The former investigators found an interaction of word frequency and visual field in two of four bilateral naming experiments. However, in one case (experiment 1) the RVF advantage was larger for low-frequency words, while in the other (experiment 3) a greater RVF advantage was found for high-frequency words. Heister's results are equally equivocal. In lexical decisions to German CVC (consonant-vowel-consonant) strings she found a larger RVF advantage for highly familiar words (for reaction time, but not accuracy, Heister, 1983; for accuracy, but not reaction time, Heister et al., 1983), which was not replicated in a subsequent study with the same stimuli (Heister, 1984). In each of these experiments error rates for low frequency words were very high (in both visual fields), so floor effects may have accounted for the diminished asymmetry. In any case, the unreplicability of these results casts doubt on their veracity, and they do not present a serious challenge to the position that word frequency has comparable effects across the visual fields.

It has also been claimed that age of acquisition, independent of word frequency, affects lexical access: words acquired earlier in life being more readily accessible than those learned later (Gilhooly & Gilhooly, 1979). The results at VHF experiments examining this variable parallel those obtained for word frequency: equivalent RVF advantages are found for early- and later-acquired words regardless of whether acquistion age is defined in terms of spoken (Ellis & Young, 1977; Young & Bion, 1980; Boles, Rogers, & Wymer, 1982) or written (Young, Bion, & Ellis, 1982) language norms.

To summarize, though it appears that the hemispheres rely on fundamentally different encoding methods, once this has occurred access to lexical representations proceeds in a comparable fashion whatever the field of input. Low-frequency words suffer no greater disadvantage when presented to the right hemisphere than do more familiar words. This may mean that strings input to the LVF and RVF ultimately address similar or identical lexical stores via similar mechanisms; it is also consistent with the view that somewhat separable lexicons are available to each hemisphere, but that these are addressed by comparable frequency-sensitive mechanisms. We now consider evidence that can potentially differentiate between these two possibilities—that is, contrasts between words of equivalent frequency but different syntactic or semantic classes.

Word Class Effects

One of the most intriguing questions regarding hemispheric processing of language is whether there is some syntactic or semantic subset of the lexicon that is directly accessible to the right hemisphere. Nonlateralized word recognition studies (with normal subjects) provide little evidence of differential lexical access for words of different semantic or syntactic classes (see Gordon & Caramazza, 1982, 1985). Thus, if such a dissociation could be demonstrated neuropsychologically (i.e., reliable visual field–word class interactions), this would entail revision of current information processing models of the lexicon.

Syntactic Class

Most VHF lexical studies either use nouns only or do not analyze the data according to grammatical class, and some studies that report results by visual field and syntactic class do not statistically test the critical visual field–word class interaction (Hines, 1976; Day, 1979). Though the data are sketchy, lexical decision (Shanon, 1979a, 1979b; Day, 1979) and naming (Hines, 1976) studies provide little support for the view that nouns, adjectives, and verbs differ in their hemispheric accessibility.

In contrast, Bradley and Garrett (1983) found different visual field effects for content (nouns, verbs, adjectives) and function (auxiliaries, pronouns, etc.) words in a naming task. In the RVF, content items were recognized more accurately than function terms, while there was no difference in response to the two categories in the LVF. Bradley and Garrett argue that function (closed class) words can be accessed by a specialized process that is only available to the left hemisphere. This is quite a strong claim, and at present there is little additional data to support it. Saffran et al. (1980) examined laterality effects for function (closed class) words and several classes of nouns in a naming paradigm, but found no interaction of visual field and word type. However, their study was not designed as a critical test of the open–closed class distinction, and it is unclear whether the nouns and functors were matched for frequency. Chiarello and Nuding (1987) did control for word frequency, but were unable to replicate the data of Bradley and Garrett in either naming or lexical decision tasks. In fact they found a small, but statistically significant, effect in the opposite direction: there was no differential processing of content and function words in the RVF, while in the LVF, function words were processed less efficiently than content items.

Clearly, consistent visual field differences in the processing of content and function words have not been demonstrated. Chiarello and Nuding (1987) suggested that the global content–function word dichotomy may not adequately capture lexical processing differences across the hemispheres. Many subtle semantic–syntactic distinctions are only partially subsumed under the content–function dichotomy (see Goodglass & Menn, 1985), and these may have varied in the above-mentioned studies.

Two investigations compared laterality effects for simple and agentive nouns (the latter are those derived from verbs, such as "worker"). Surprisingly, Caplan et al. (1974) found that agentives were named more accurately than simple nouns matched for frequency, and this effect was equivalent across the visual fields (the RVF advantage did not differ for the two classes). Goodall (1984), measuring vocal naming latencies, found no visual field asymmetry for simple nouns, but a RVF advantage for the agentives (simple and agentive words were named equally fast in the RVF, while in the LVF agentive nouns yielded slower responses than simple nouns). This may indicate a selective LVF–RH deficit for recognizing morphologically complex words, but methodological considerations suggest caution. The simple nouns used in this study were, on average, more concrete than the agentive words. Goodall attempted to rule out an explanation based on concreteness by having subjects pluralize the same simple nouns presented to either visual field. Under these conditions, a RVF advantage was obtained. Unfortunately, this does not really speak to the issue because the *inflectional* process of pluralization is not necessarily comparable to the process of reading a *derivationally* complex form. Given the negative findings of Caplan et al. (1974) it is premature to conclude that morphologically complex words are less accessible to the right hemisphere than simpler forms.

More well-controlled studies examining grammatical class variables are certainly needed. At present, the most obvious distinctions (i.e., noun vs. verb, content vs. function word) do not appear to map onto consistent hemisphere differences in word recognition. This may simply reflect the fact that *lexical access* operates without regard to grammatical class distinctions, whatever the hemisphere of input.

Semantic Class

The search for differential laterality effects for the recognition of concrete and abstract words[6] has become something of a growth industry in neuropsychology; more studies have examined this than any other lexical variable. This collective research investment has yet to pay off. The number of studies finding a greater left hemisphere asymmetry for abstract than for concrete items is just large enough to whet the appetite (Ellis & Shephard, 1974; Hines, 1976, 1977; Day, 1977, 1979; Elman, Takahashi, & Tohsaku, 1981; Bub, 1982, experiment 3; Young & Ellis, 1985, experiment 5). However, the number of marginal findings (Bradshaw & Gates, 1978, experiment 3; Hines, 1978; Moscovitch, 1981) or negative findings (Orenstein & Meighan, 1976; Hatta, 1977a; Saffran et al., 1980; Schmuller & Goodman, 1979; Shanon, 1979a, 1979b; Bradshaw et al., 1981a; Boles, 1983; Lambert & Beaumont, 1983) is sufficient to induce dyspepsia. It is tempting to attribute this state of affairs to differences in "procedural variables."

Certainly these studies span the range of tasks, viewing conditions, and response measures employed in lexical lateralization research. However, this is no less true of the studies reviewed above, where reliable hemispheric effects were obtained for many variables (e.g., exposure time, pattern masking, word length, frequency), despite substantial changes in procedure across studies. If the variability in the abstract–concrete findings can indeed be tied to specific experimental conditions, we must still acknowledge that the effect is much more elusive than other factors known to affect VHF word recognition.

In any case this point is moot because there is no set of experimental conditions that can be used to predict whether the visual field–concreteness interaction will occur, even on a post-hoc basis. Positive and negative results have been reported both with masking (Marcel & Patterson, 1978 vs Lambert & Beaumont, 1983) and without (Hines, 1976 vs. Boles, 1983), with horizontal (Ellis & Shepherd, 1974 vs. Schmuller & Goodman, 1979) and vertical (Day, 1977, 1979 vs. Shanon, 1979a, 1979b) orientation, unilateral (Hines, 1976 vs. Lambert & Beaumont, 1983) and bilateral (Young

[6] Because the imageability and concreteness dimensions are so highly correlated, concrete–abstract are used as cover terms for either dimension.

& Ellis, 1985 vs. Lambert & Beaumont, 1983) presentation, with high (Hines, 1977 vs. Bradshaw et al., 1981a) and low (Ellis & Shepherd, 1974 vs. Orenstein & Meighan, 1976) statistical power, for lexical decision (Day, 1977, 1979 vs. Bradshaw et al., 1981a) as well as naming (Hines, 1976, 1977 vs. Hines, 1978; Boles, 1983), and for Kanji script (Elman et al., 1981 vs. Hatta, 1977a). In an earlier review Boles (1983) concluded that procedural variation could not account for the lability of the visual field–concreteness interaction. Results reported since his review have only supported this conclusion. For example, the suggestion of Lambert & Beaumont (1983) that the interaction is an artifact of failure to control order of report strategies cannot account for the positive findings of Young and Ellis (1985).

Perhaps at this juncture the initial question should be re-examined. Two proposals have been offered for the putative hemispheric dissociation between abstract and concrete words. The first argues that at least partially separate lexicons are accessed by each hemisphere and that the words available to (or represented in) the right hemisphere lexicon are predominantly concrete and imageable (Coltheart, 1980a). The second hypothesizes that the RH recognizes words by a special mechanism that is inadequate for abstract, nonimageable terms: "It is possible that the pattern recognition system for words arriving initially at the right hemisphere is facilitated by imagery" (Ellis & Shepherd, 1974, p. 1036); "direct visual access via an immediately generated image" (Bradshaw, 1980, p. 180). Presumably concrete words presented to the LVF–RH would immediately evoke an image that is used to access semantic representations needed for comprehension. Because abstract words cannot be imaged, the RH mechanism could not support their recognition. This is not just a claim that the right hemisphere may be better than the left at constructing images (Paivio & te Linde, 1982), but a stronger argument that images mediate access to semantics in the RH, and thus precede rather than follow comprehension. Both proposals, then, postulate a hemispheric difference at the stage of lexical (or semantic) access. This contrasts with the dual code theory of Paivio (Paivio & O'Neill, 1970; Paivio, 1971), which posits a role for imagery only after (concrete) words are recognized.

We have seen that prelexical processes, which culminate in an abstract graphemic representation, seem to be independently carried out in the hemisphere of input. Thus each hemisphere has available to it a representation with which to access the lexicon. If the RH were to use this representation to access its own abridged lexicon, or to construct an image, reliable visual field–concreteness interactions would be demonstrable in naming and lexical decision tasks. This seems not to be the case, which argues against hemisphere differences in lexical access attributable to imageability. However, this does not argue against the possibility that hemisphere differences in processing abstract and concrete words may arise postlexically (i.e., after semantic information is retrieved).

In fact in nonlateralized studies with normal readers, word concreteness has little effect on performance in simple reading tasks. In naming, response latency and recognition thresholds do not differ for abstract and concrete words matched for frequency (Winnick & Kressel, 1965; Paivio & O'Neill, 1970; Richardson, 1976; but cf. Riegel & Riegel, 1961). In lexical decision, Richardson (1976) found no effects of concreteness or imageability, while James (1975) reported no effect of concreteness on reaction times to high-frequency words, but an advantage of concrete over abstract words for low-frequency nouns. Kroll and Merves (1986), however, found no consistent interaction of frequency and concreteness in a series of lexical decision experiments (cf. Gernsbacher, 1984). In addition, they found only marginal effects of concreteness in all but one condition: where the abstract and concrete words were blocked and concrete blocks occurred first. Only under these conditions were concrete words responded to faster than abstract words. Kroll and Merves note that their results are problematic for the view that concreteness influences the lexical access process.

These findings are mentioned for two reasons. First, to point out that there would be little reason from the data on normal word recognition to expect concreteness to affect lexical access. Dissocations between abstract and concrete words are typical only in memory tasks or in those that require substantial postlexical processing (Paivio, 1971). Second, to note that the lability in visual field effects for concrete versus abstract words parallels the only occasional finding of concreteness effects in nonlateralized word recognition experiments. This suggests that the unreplicability of the VHF studies may stem from the nature of the tasks used to dissociate processing of abstract and concrete words. Semantic variables such as concreteness may only play a role subsequent to lexical access, when semantic information becomes available for further processing, regardless of the hemisphere of input. If this is the case, tasks such as lexical decision and naming that do not require semantic retrieval are not the best tests of possible hemispheric differences in the processing of abstract versus concrete words. It is tempting to speculate that subjects may optionally retrieve semantics in these tasks, and that this may account for occasional positive findings. But even here the locus of the effect would be postlexical. The role of each hemisphere in postlexical semantic processing will be considered in a subsequent section.

To sum up these findings, there is to date no convincing data to suggest that certain classes of words are any less accessible than other word types when presented to the right hemisphere. There is little or no evidence that abstract, low-frequency, or function words show a greater disadvantage when input to the LVF–RH than do concrete, high-frequency, or content words—at least in simple word recognition tasks. Thus there is little support for the view that unequal lexicons are available to each hemisphere; rather, the current data suggest that in the normal brain each hemisphere has access to the same lexical information. However, it remains to be determined whether this information is redundantly represented in each hemisphere or whether both hemispheres access the same lexical store.

Automatic Lexical Priming

In tasks such as lexical decision and naming, performance is facilitated when the word to be recognized is preceded by a semantically related (prime) word (Meyer & Schvaneveldt, 1971). It appears that there are two sources for such priming effects: (*a*) an automatic process that operates within the lexicon, facilitating lexical access for semantically related words, and (*b*) a second cognitively controlled process whose primary locus is postlexical (Neely, 1977; Seidenberg et al., 1984). Automatic priming is thought to occur as function of associative connections that exist between related words in lexical memory: when the prime word is presented, its memory representation is activated, this activation spreads via links in the semantic network to related representations, and temporarily lowers their recognition thresholds (Collins & Loftus, 1975). Priming that is automatic does not depend on the subject's attention to (or awareness of) the prime word, and can be demonstrated by masking the prime word (Fowler, Wolford, Slade, & Tassinary, 1981), by including only a few related primes in the experiment (low-probability priming) (Fischler, 1977), or by employing very brief prime–target SOAs (stimulus onset asynchronies) (Fischler & Goodman, 1978).

Automatic priming can thus be used as a technique to "trace out" the lexical connections between words in either hemisphere. When prime words are masked to prevent their awareness, there is equivalent semantic priming across the visual fields (Marcel & Patterson, 1978; Walker & Ceci, 1985, experiment 2). Chiarello (1985, experiment 6), under low-probability priming conditions, found greater reaction time priming in the LVF (priming measured in error rates did not differ across the visual fields).[7] However, this effect was not replicated in a subsequent low-probability semantic priming experiment in which equal priming was obtained in each visual field (Chiarello, Senehi, & Nuding, 1987). Eglin (1987) has also reported equivalent visual field semantic priming under automatic processing conditions.

It is clear from these results that there is no RVF advantage for the automatic semantic priming effect. Why such effects should sometimes be larger in the LVF is puzzling. Because the LVF priming advantage is quite

[7] Zaidel (1983a) and Walker and Ceci (1985, experiment 1) have also reported larger semantic priming in the LVF, but only for one response measure. However, the experimental paradigms used in these investigations do not allow us to differentiate automatic from nonautomatic priming (cf. Neely, 1977). Thus we cannot determine whether the observed priming effects were strictly automatic.

labile, it is perhaps best not to draw any firm conclusions until we can better identify the conditions under which it occurs. However, because automatic semantic priming is certainly not larger in the RVF, it is unlikely that the overall RVF advantage for recognizing words can be attributed to hemisphere differences in the spread of activation through the semantic network. To date, the automatic priming results do not invalidate the position that both hemispheres access similar semantic information.

Several priming studies have investigated the automatic activation of different types of semantic information. Both concrete and abstract primes produce equivalent facilitation in each visual field (Marcel & Patterson, 1978; Chiarello et al. 1987, experiment 3), thereby supporting the view of equivalent lexical access for concrete and abstract words in each hemisphere. Burgess and Simpson (in press) measured the automatic activation of ambiguous word meanings at a very brief (35 msec) prime-target SOA. Under these conditions greater priming was obtained for dominant than for subordinate meanings in each visual field, although the data suggested a slightly slower build-up of activation for subordinate meanings in the LVF. The latter finding deserves further scrutiny, but at present the collective semantic priming results imply that semantic spreading activation operates in a similar fashion in each hemisphere across all word types.

Lexical priming can also be obtained when the prime and target words are orthographically or phonetically related (Hillinger, 1980; Evett & Humphreys, 1981). In an automatic (low-probability) priming experiment Chiarello (1985) found priming between orthographically related pairs (e.g. BEAK-BEAR) to be present in the LVF, but not in the RVF (experiment 4). The converse occurred for phonetically related words (e.g., MOOSE-JUICE); here priming was obtained in the RVF, but not in the LVF (experiment 5). If these findings can be replicated, it would suggest that orthographic "links" between words are accessible only to the right hemisphere, and phonological "links" only to the left hemisphere. Forster (1978) proposed a lexical model that would allocate such priming to spreading activation within separate "peripheral access files" that are organized by orthographic or phonetic features. These files are distinct from, and mediate access to, the semantic lexical system. Under this interpretation, nonsemantic priming (which appears to be asymmetrical) has a different lexical basis than semantic priming (which has a much more symmetrical effect) (for discussion, see Chiarello, 1985).

Summary

Although many prelexical processes are qualitatively different in each hemisphere, there is little compelling evidence that unequal lexicons are available to the normal right and left hemispheres. The view that the right hemisphere accesses an "abridged dictionary," consisting of high-frequency, concrete words, is not supported by the current VHF findings. Word frequency has an equivalent effect across the visual fields, and reliable shifts in asymmetry attributable to word class (semantic or syntactic) have yet to be demonstrated. It is possible that orthographic and phonetic relations between words are differentially represented in each hemisphere, but a great deal more research is needed in this area.

Although it appears that each hemisphere accesses a similar semantic store, this does not necessarily imply a left hemisphere locus. We cannot rule out the possibility of bihemispheric representation. (It is of course an extremely thorny problem to attempt to "locate" a memory store within the brain.) The hemisphere differences in prelexical encoding processes could account for the overall RVF advantage even assuming hemispheric equivalence in accessing lexical information. Thus it cannot be proven that graphemic representations derived from LVF input require interhemispheric transfer in order to access semantic information.

Postlexical Processes

This section considers the hemispheric basis of processes occurring subsequent to lexical access, which depend on information retrieved from the lexicon.

The Lexical Decision

Can either hemisphere make lexical decisions? While a RVF advantage for this task is generally found (see Leiber, 1976; Barry, 1981; Chiarello et al., 1984), this could reflect asymmetries in encoding, in lexical access, or in the postlexical decision stage. Results reviewed here certainly suggest that encoding contributes to the overall RVF effect, but there is also evidence that lexical decision criteria differ across the visual fields. Signal detection analyses find robust visual field differences in sensitivity (d'), and smaller, but statistically significant, differences in response criteria (β) (Chiarello et al., 1984, 1986, in press; Babkoff, Genser, Sing, Thorne, & Hegge, 1985). The latter occurs due to a more stringent criterion (nonword response bias) in the LVF, and a more lax criterion (word bias) in the RVF.

The response criteria adopted by the subject should not affect stimulus encoding nor lexical access processes. Thus, these criteria will reflect biases operating during the decision process after lexical access, when the final word–nonword decision is determined (Chiarello et al., in press; see Figure 2.1). The findings of differential visual field response criteria thus imply that the outcome of the lexical access process (whether or not a lexical representation has been found) is differentially evaluated by each hemisphere when the final lexical decision is made. In other words, in addition to the hemisphere differences in prelexical (encoding) operations, the hemispheres also differ in the postlexical processes recruited for lexical decision. This supports the view that either hemisphere can make lexical decisions, abeit with differing efficiency and using different response criteria.

A related, and puzzling, finding is the contrast in visual field performance for word and nonword responses in the lexical decision task. Though most investigators find a RVF advantage for word decisions, nonword responses are typically equivalent (in both accuracy and reaction time) across the visual fields (Leiber, 1976; Bradshaw & Gates, 1978; Babkoff & Ben-Uriah, 1983; Brand et al., 1983; Chiarello et al., 1984, 1986; Chiarello, 1985; Hardyck et al., 1985a, 1985b). A few studies report a RVF advantage for nonwords (Barry, 1981; McKeever & Hoff, 1982; Heister, 1984), yet procedurally there is little to distinguish these studies from those finding no nonword asymmetry. As Chiarello et al. (1984) showed that the retest reliability for nonword decisions is extremely low ($r < .40$), it seems reasonable to conclude that there is no reliable visual field asymmetry for nonwords comparable to that found for word decisions.

The word–nonword discrepancy cannot be accounted for by postulating that lexical access is more efficient when initiated by the left hemisphere because this should lead to more rapid or accurate word and nonword decisions for RVF strings (assuming that an exhaustive lexical search is made prior to a nonword response; cf. Forster, 1976). However, under tachistoscopic conditions an exhaustive lexical search may not be undertaken. The alternative would require a response deadline, in which a nonword response will be made if no lexical entry has been accessed when the deadline elapses. If similar response deadlines are adopted for either visual field, one would expect equivalent visual field performance for nonwords. Assuming a slower onset of lexical search in the LVF (due to a longer stimulus encoding process), lexical entries for words in the LVF would be accessed later, and in some cases not accessed at all, prior to the response deadline, producing slower and less accurate word responses than in the RVF.

Under this interpretation, words presented to the LVF would be less likely to achieve lexical access than those presented to the RVF. Yet this would be attributed to an asymmetry at an earlier processing stage and not to hemisphere differences in the lexical access process. Slower onset of lexical search for strings presented to the LVF could also be attributed to interhemispheric transfer, rather than longer stimulus encoding within the RH. However, it appears that visual field differences for lexical decisions are unlikely to occur under conditions that minimize hemisphere differences in encoding (i.e., vertical orientation) (Day, 1977, 1979; Shanon, 1979a, 1979b).

Although the above interpretations are speculative, the lexical decision findings converge on the view that the asymmetry for this task arises both pre- and postlexically, with no compelling evidence for qualitatively different lexical access procedures. However, because the outcome of the lexical access process is evaluated via different response criteria in each hemisphere, this provides evidence that the lexical decision is reached in the hemisphere of input.

Pronunciation

An important question in the interpretation of lexical laterality effects is the extent to which pronunciation contributes to the nearly ubiquitous left hemisphere advantage. It is clear that the requirement for a discriminative vocal response is neither a necessary nor a sufficient condition for the RVF advantage in verbal tasks. In lexical decision, the RVF advantage occurs in the absence of vocalization, regardless of responding hand (Babkoff & Ben-Uriah, 1983; Chiarello et al., 1986), which suggests that the asymmetry does not depend on which hemisphere programs the final motor output. In addition, a LVF advantage can occur even when pronunciation is required (Gibson, Dimond, & Gazzaniga, 1972; Bryden & Allard, 1976; Hatta, 1977b). Thus, to determine the contribution of pronunciation to the overall visual field advantage we must consider word recognition studies that contrast verbal and nonverbal responses, but with all other experimental variables held constant.

Only one lateralization study appears to meet these criteria. Klein, Moscovitch, and Vigna (1976) found that when four-letter words were named, approximately 80% of the correct responses were RVF reports. But when subjects reported the same lateralized words by pointing (using an array of targets and distractors in free vision), only 60% of the correct responses were from the RVF. This suggests that overt pronunciation may account for a great deal of the asymmetry typically seen in naming tasks. However, because the vocal response required free report, but the nonverbal response forced choice recognition, this interpretation remains speculative.

More indirect evidence can be obtained by comparing lexical decision and naming asymmetries for the same words. Because in these tasks words are assumed to be processed similarly up to the point of lexical access, differential task asymmetries would implicate a divergence in postlexical operations. If a larger RVF advantage were obtained with naming, this would suggest that the pronunciation stage is more dependent on LH processes than the (nonvocal) decision stage in lexical decision. Unfortunately, simple task comparisons have yielded inconsistent results with one study finding a greater RVF advantage for naming (Bradshaw & Gates, 1978, experiment 4; Bradshaw, Gates, & Nettleton, 1977, experiment 1), another reporting a greater RVF advantage for lexical decision (Healey, Waldstein, & Goodglass, 1985, experiment 1), and a third finding an equivalent RVF advantage across tasks (Chiarello & Nuding, 1987). However, a more recent study suggests a possible explanation for these conflicting results (Chiarello et al., in press).

Recall that there is a visual field difference in the response biases adopted in the lexical decision task. As naming does not involve a binary classification response, no such bias is present for this task (Seidenberg, 1985). However, initial studies comparing lexical decision and naming did not take this task difference into account. Chiarello et al. (in press) compared visual field asymmetries for naming and various lexical decision tasks (i.e., vocal and right-hand manual go–no-go vs. yes–no paradigms) and obtained several results relevant to this issue. First, the visual field difference in response bias (β) depended to some extent on the particular lexical decision paradigm employed. Second, an apparent difference in task asymmetries (a greater RVF advantage for yes–no lexical decision than for naming) disappeared when lexical decision response bias was statistically eliminated from the task comparisons. In addition, there was no alteration in visual field performance attributable to the requirements for vocal, as opposed to manual, response.

These findings suggest that the conflicting results for tasks comparisons obtained in previous studies may be attributable to variations in lexical decision response bias asymmetry, and not to intrinsic task differences relating to the necessity for pronunciation. This leads to the rather surprising conclusion that the process of articulating speech requires no greater access to left hemisphere operations than does a simple (right-hand) manual response. Clearly, further studies comparing vocal and manual responses for lexical tasks are needed before this conclusion can be accepted.

A related issue concerns whether or not the hemisphere asymmetry for pronunciation differs depending on the relative contributions of lexical and nonlexical information to this process (see Figure 2.1). As the nonlexical route plays a greater role in the pronunciation of nonwords than words (Rosson, 1985), word–nonword comparisons may provide relevant evidence.[8] Bradshaw and Gates (1978, experiment 4) compared the naming of four-letter words and nonwords, one-half of the latter being homophonic to real words. The RVF advantage was significantly larger for homophonic nonwords than for the other two classes (which did not differ). However, this result was not replicated in a subsequent study where the RVF advantage was constant for all three stimulus types (Bradshaw & Taylor, 1979). Young et al. (1980) compared naming of CVC nouns and nonwords in a bilateral cued order of report paradigm. Although there was a significant RVF advantage for nonwords for both first and second reports, a RVF advantage for words was obtained only on second reports. (There was no

[8] Studies in which word and nonword strings were reported letter-by-letter (Dornbush & Winnick, 1965; Axelrod, Haryadi, & Leiber, 1977) are not relevant to the pronunciation issue and are not considered here.

evidence of ceiling effects for words reported first.) The more robust asymmetry for nonwords suggests that when hemisphere differences in encoding are minimized (by using three-letter strings), a RVF–LH advantage may depend on greater use of nonlexical pronunciation codes.

Some of the variability in the above studies may be attributable to whether the words and nonwords were presented in a mixed fashion (as in the studies of Bradshaw), or in separate blocks (as in Young et al., 1980). It appears that words are more likely to be pronounced "nonlexically" when they appear among nonwords than when words and nonwords occur in separate blocks (Frederiksen & Kroll, 1976; Hudson & Bergman, 1985). Thus, future word-naming research using the blocked and mixed manipulation might provide a more direct test of whether asymmetries differ for lexical and nonlexical pronunciation.

In any case, it should be noted that a RVF advantage for pronouncing nonwords is invariably found (Bradshaw & Gates, 1978; Bradshaw & Taylor, 1979; Young et al., 1980; Young, Ellis, & Bion, 1984; Young & Ellis, 1985). These results are consistent with the view that the nonlexical pronunciation route is greatly dependent on left hemisphere processing. An experiment done by Bradshaw and Gates (1978, experiment 2), however, suggests that it is the requirement to pronounce nonwords that produces the RVF advantage rather than nonlexical print-to-sound conversion. In this study, subjects made lexical decisions to nonword letter strings, half of which were homophonic to real words (e.g., kirl). Thus, correct decisions required phonologic conversions, but not overt pronunciation. No visual field differences were observed in this task. If this finding is replicable, one could attribute the RVF nonword advantage to the pronunciation process per se, rather than to nonlexical print-to-sound transcoding.

A final consideration is the hemispheric locus of the spelling-to-sound regularity effect: words having regular spellings are typically named faster and more accurately than those with exceptional spellings. The precise mechanism responsible for this effect is in doubt, but it appears that it is attributable to the presence of conflicting information about pronunciation of exception words (due to a conflict in codes derived lexically and nonlexically, or a conflict between several different lexically activated codes) (Parkin, 1984; Seidenberg, Waters, Barnes, & Tanenhaus, 1984). Under any interpretation, some postlexical process would be required to determine the appropriate pronunciation for exception words. Parkin and West (1985) found a spelling regularity effect for naming words shown in the RVF, but not in the LVF, and interpret this as evidence that the right hemisphere cannot convert print into a phonological code prior to pronunciation. They rule out the possibility that pronunciation is determined in the left hemisphere for LVF stimuli because this would have resulted in regularity effects in both visual fields. Presumably, pronunciation is determined nonphonologically in the right hemisphere, with this information transferred to the left hemisphere for the final motor output. This is quite an extreme position, and requires confirmatory findings. However, there is some reason to question these results. Low-frequency, vertically arrayed words were used in this experiment, resulting in an unusually low level of performance, particularly for LVF stimuli. Thus the absence of a regularity effect in the LVF could be a consequence of floor effects, rather than a true hemisphere difference.

It is generally accepted that discriminative vocal responses are highly lateralized to the left hemisphere, based primarily on the evidence from commissurotomy patients (e.g., Sperry, Gazzaniga, & Bogen, 1969). The data from normal subjects do not refute this claim, but the relevant evidence is scarce at best. It is still unclear to what extent the robust RVF advantage for naming is attributable to left hemisphere control of pronunciation per se (independent of the demonstrable visual field differences in prelexical encoding processes), or to relative reliance on nonlexical versus lexical pronunciation codes. The current evidence is barely suggestive of a greater LH role in pronunciation, particularly when this is nonlexically derived. There is a critical need for future lateralization studies with neurologically normal individuals to examine the vocal–nonvocal dichotomy, and the derivation of pronunciation codes.

Semantic Judgments

Postlexical semantic processing is required in tasks where the response must be based on (or can be influenced by) semantic information retrieved from the lexicon. Several VHF studies have examined semantic categorization judgments. When two words are presented and subjects decide whether the words are members of the same semantic category (e.g., CHAIR–TABLE) a RVF advantage is found (Gross, 1972; Urcuioli, Klein, & Day, 1981; Klein & Smith, 1985). However, it appears that this asymmetry only emerges after subjects have had considerable experience with the same semantic categories, although not necessarily the same words from these categories (Urcuioli et al., 1981; Klein & Smith, 1985). Thus, the RVF advantage may depend on repeated retrieval from the same set of semantic categories.

Comparable laterality findings have been reported in a controlled (i.e., high probability, nonautomatic) semantic priming paradigm, where "related" prime and target words were members of the same category (Chiarello, 1985, experiment 3). In this lexical decision experiment the priming effect was substantially larger for words presented to the RVF–LH. Current research suggests that such priming (in contrast to the automatic, spreading activation priming discussed earlier) involves a postlexical semantic integration stage. In this stage the semantic relationship between the prime and target words is examined, and contributes evidence to the lexical decision mechanism (de Groot, 1984; Seidenberg, et al. 1984). Thus although controlled semantic priming is ostensibly quite different from same/different category judgments, both may have involved a similar process of postlexical category matching. The above studies suggest that this process is more efficiently performed when stimuli are input to the left hemisphere, even though only concrete words were employed.

Contrasting results are found in a slightly different categorization task. There are no visual field differences when judging whether a word is a member of a particular semantic category (e.g., FURNITURE–TABLE), as long as the category is concrete (Day, 1977; Urcuioli et al., 1981). However, a RVF advantage was found for similar judgments about abstract categories (Day, 1977, experiment 2). Considering only the results for concrete words, it appears that the process of category matching (where superordinates of two words must be retrieved and compared) requires greater left hemisphere involvement than does a judgment of category membership (where only one superordinate must be retrieved).

An abstract–concrete dissociation for semantic judgments similar to that of Day (1977) was also observed by Rodel, Dudley, & Bourdeau (1983). They found a RVF advantage for synonymy judgments about abstract words, but no visual field difference when concrete words were judged: there was no difference between abstract and concrete words in the RVF, while in the LVF judgments about abstract words were considerably slower than concrete words. We have found complementary results in our laboratory using a high-probability (controlled) semantic priming paradigm (Chiarello et al., 1987). We compared priming effects for words primed with either abstract or concrete terms (HUNGER–FOOD vs. BREAD–FOOD). Equivalent effects were obtained for abstract and concrete primes in the RVF, while abstract priming was considerably less effective than concrete priming in the LVF. In other words, there was a RVF advantage in the size of the abstract, but not the concrete, controlled priming effect. If it is assumed that such effects are predominantly postlexical (de Groot, 1984; Seidenberg, 1985), the data imply a LVF–RH deficit in processing of abstract words which occurs subsequent to lexical retrieval.

Although only a few studies are relevant to the issue of hemisphere differences in postlexical semantic processing, the current evidence suggests two conditions under which a RVF–LH advantage may occur: when words are matched on the basis of their semantic category membership, or when processing depends on the retrieved meaning of abstract words. No doubt there are other conditions, yet to be examined, which would also elicit left hemisphere predominance. However, the absence of visual field differences in several semantic judgment experiments suggests that at least some types of postlexical semantic processing are performed with equivalent efficiency in either hemisphere.

Visual and Phonetic Judgments

Matching words based on visual identity need not require lexical processing, yet in such tasks words are matched faster than legal nonwords, and word frequency effects are present (Chambers & Forster, 1975; Brand et al. 1983). This suggests that lexical access is involved, and that lexically retrieved information may be used in the comparison process (see discussion in Chambers & Forster, 1975). Thus visual field word matching studies may be more indicative of postlexical than prelexical visual processes. Visual identity matching for simultaneously presented words yields no differences between visual fields (Sasanuma, Itoh, Kobayashi, & Mori, 1980; Brand, et al. 1983) and Sasanuma et al. found the visual field equivalence for both Kana and Kanji words. Gibson et al. (1972) used a 500-ms separation between presentation of the two words and reported a LVF advantage for visual identity judgments. Chiarello (1985, experiment 1) obtained a related finding for controlled (high probability) priming using visually similar words (BEAK–BEAR). A 500-ms interval separated presentation of the prime and target words, and greater priming was obtained for pairs in the LVF.

Although a great deal more research is needed, these results suggest that both hemispheres are equally adept at simultaneous comparisons of visual lexical information, but that the right hemisphere may be better able to retain lexical information in a visual code when delayed comparisons are required. If lexically retrieved information is used in these tasks, the visual field equivalence in the simultaneous condition implies that visual orthographic information is retrieved with equal speed and accuracy in either hemisphere.

Even fewer studies have examined lexical judgments based on phonetic comparisons. Sasanuma et al. (1980) examined rhyme decisions for Kana words and homophone judgments for Kanji words and obtained a RVF advantage in both cases. Likewise Rodel et al. (1983) found a RVF advantage for judgments of homophony. Because lexical judgments do not invariably produce RVF advantages, these findings suggest that phonetic comparisons of lexically retrieved information are more efficiently performed by the LH. However, Chiarello (1985, experiment 2) found that controlled (high probability) priming with visually dissimilar rhymes had an equal effect in each visual field. If such priming involves a postlexical comparison stage, this would provide counter evidence to the claim that postlexical phonetic processing is invariably asymmetrical.

Summary

The role of post-access processes in word recognition tasks has, until quite recently, been unexplored. It is not surprising, then, that there are few relevant VHF studies. Perhaps the only generalization that can be made at this point is that both the nature of the information retrieved, and the postaccess processes required, will determine the observed visual field asymmetry.

There is little reason to assume that all postlexical processing requires left hemisphere involvement. The qualitatively different response biases in the LVF and RVF for lexical decision suggest that postlexical processing occurs in each hemisphere via different criteria. Thus either hemisphere may perform the lexical decision. In addition, performance of at least some visual and semantic judgments seems not to vary with the hemisphere of input. Yet there are intriguing leads that some postlexical semantic processing is specific to the left hemisphere. Because there is little evidence for hemisphere asymmetry in access to semantic information, further delineation of the left hemisphere's role in postlexical semantics is needed.

It is of interest that evidence for left hemisphere control of the pronunciation process is not abundant. This position can be neither refuted nor strongly supported by current visual lateralization research with normal individuals. Use of experimental paradigms which permit separation of encoding and lexical access from the production process in naming (Balota & Chumbley, 1985) would be a welcome addition to the techniques currently used in VHF lexical research.

It is important to note that findings of differential hemispheric performance for postlexical operations (such as the differing lexical decision response bias) imply independent hemispheric processing for earlier (prelexical and lexical) processes. This would argue against a unilateral left hemisphere locus for the semantic store, and in favor of a bihemispheric, redundant representation of semantic information.

Conclusions

Review of current VHF findings reveals some important methodological considerations for the design of future lexical studies. Some variables (exposure time, unstructured masking, word frequency) have an equivalent effect on performance in each visual field, and thus can be considered hemispherically "neutral." Others, such as orientation, pattern masking, and word length have differential hemispheric effects, and thus cannot be treated as background variables devoid of theoretical interest. The use of vertically oriented words, in addition to lowering overall accuracy (often dramatically), may prohibit "whole word" encoding procedures normally employed by the left hemisphere. Under most conditions, pattern masking will disrupt perceptual encoding to a greater extent in the LVF than in the RVF, unless the target–mask interval is independently adjusted for each visual field. The magnitude of the RVF advantage increases with word length, due to an increasing load on LVF serial encoding processes. Though there is in general no "correct" set of presentation conditions for VHF lexical studies, the parameters chosen should be motivated by the experimenter's desire to diminish or enhance hemisphere asymmetries in prelexical encoding. At the very least, comparisons between ostensibly similar studies should take these factors into account.

Do the normal left and right hemispheres have distinctly different reading profiles? I believe the answer is yes, although we are far from understanding many aspects of lexical processing in the normal brain. Words presented to the left hemisphere are encoded by a rapid automatic process specialized for the perception of words. Lexical codes become available almost immediately and assist in the recognition of component letters. In contrast, words input to the right hemisphere are encoded by a slower, serial process that becomes increasingly inefficient as word length increases. This right hemisphere mechanism encodes all letter strings in a similar fashion, with lexical codes available only after each letter has been separately processed. Thus the RH does not use encoding mechanisms specialized for recognizing words, relying instead on more general purpose processes. These are typically less efficient than those of the LH, but may be favored when words are presented in unusual visual formats that cannot engage lexically specialized LH processors. This hemispheric differentiation in encoding allows for considerable flexibility in recognizing words under a wide variety of conditions.

Despite substantial differences in prelexical operations, each hemisphere appears to access similar semantic information. The current evidence does not support the view that the RH accesses an abridged lexicon, or that any particular class of word is inaccessible to the normal right hemisphere. Despite the pronounced hemispheric differences in prelexical encoding, there is little evidence that the process of lexical access is any less efficacious when initiated by the RH. Nor can an asymmetrical locus for the semantic store be established. At present, then, left and right hemisphere reading profiles cannot be distinguished on the basis of access to lexical semantic information.

The hemispheric contributions to postlexical processing vary considerably with the task and the type of information retrieved from the lexicon. There is no evidence to contradict the notion that the LH is responsible for programming the final motor output for speech, although there is remarkably little relevant data. However, it cannot yet be ascertained at what stage in the pronunciation process the LH is required. Several possibilities exist: the retrieval of lexical pronunciation codes, access to a nonlexical pronunciation route, a stage in which various codes are evaluated to determine the final pronunciation, or the programming of the actual articulatory commands. Asymmetries for postlexical processes not involving vocalization depend on the type of information retrieved and the nature of the subsequent processing. At a minimum, the RH has the capability for postlexical processing of some semantic and visual–orthographic information. Nevertheless, it appears that hemispheric differences in semantic processing, when they occur, are most likely to arise postlexically.

In sum, the reading profiles of the normal left and right hemispheres differ primarily in processes recruited prior to lexical access, and to a lesser extent in postlexical operations, with little apparent difference in lexical access or in the availability of semantic information. The normal RH reading profile bears no striking resemblence to any particular acquired dyslexia syndrome. This is not surprising, as it is unlikely that a lesion would abolish LH reading substrates, yet leave RH mechanisms completely unaffected. However, some aspects of normal RH word recognition, such as letter-by-letter reading, are obvious in the behavior of individual dyslexic patients (Patterson & Kay, 1982). Perhaps the most workable approach to comparing data from normal subjects and patients with dyslexia would be to focus on the symptom—rather than the syndrome—level of analysis. A given patient may rely to a greater than normal extent on some RH processes, but not others, depending on the nature and severity of the LH deficit. Conjoint use of VHF and information processing techniques can provide the critical data to enable us to attribute particular reading processes to the right hemisphere in both normal and reading disabled populations.

Acknowledgments. This research was supported by BSRG 2 S07 RR077068–19. I am grateful to Hiram Brownell, Susan Garnsey, Curtis Hardyck, Larry Lewandowski, Morris Moscovitch, and Andy Young for providing helpful comments on an earlier version of this chapter.

References

Axelrod, S., Haryadi, T., & Leiber, L. (1977). Oral report of words and word approximations presented to the left or right visual field. *Brain and Language*, *4*, 550–557.

Babkoff, H., & Ben-Uriah, Y. (1983). Lexical decision time as a function of visual field and stimulus probability. *Cortex*, *19*, 13–30.

Babkoff, H., Genser, S., & Hegge, F.W. (1985). Lexical decision, parafoveal eccentricity, and visual hemifield. *Cortex*, *21*, 581–593.

Babkoff, H., Genser, S.G., Sing, H.C., Thorne, D.R., & Hegge, F.W. (1985). The effects of progressive sleep loss on a lexical decision task. Response lapses and response accuracy. *Behavior Research Methods, Instruments, & Computers*, *17*, 614–622.

Balota, D.A., & Chumbley, J.I. (1984). Are lexical decisions a good measure of lexical access? The role of word frequency in the neglected decision stage. *Journal of Experimental Psychology: Human Perception and Performance*, *10*, 340–357.

Balota, D.A., & Chumbley, J.I. (1985). The locus of word-frequency effects in the pronunciation task: Lexical access and/or production? *Journal of Memory and Language*, *24*, 89–106.

Banks, W.P., Oka, E., & Shugarman, S. (1981). Recoding of printed words in internal speech: Does recoding come before lexical access? In O.J.L. Tzeng & H. Singer (Eds.), *Perception of print: Reading research in experimental psychology* (pp. 137–170). Hillsdale, NJ: Lawrence Erlbaum Associates.

Baron, J. (1978). The word-superiority effect: Perceptual learning from reading. In W.K. Estes (Ed.), *Handbook of learning and cognitive processes*, *Volume 6*

(pp. 131–166). Hillsdale, NJ: Lawrence Erlbaum Associates.
Barry, C. (1981). Hemispheric asymmetry in lexical access and phonological encoding. *Neuropsychologia*, *19*, 473–478.
Barton, M.I., Goodglass, H., & Shai, A. (1965). Differential recognition of tachistoscopically presented English and Hebrew words in right and left visual fields. *Perceptual and Motor Skills*, *21*, 431–437.
Becker, C.A. (1976). Allocation of attention during visual word recognition. *Journal of Experimental Psychology: Human Perception and Performance*, *2*, 556–566.
Besner, D. (1983). Deep dyslexia and the right hemisphere hypothesis. Evidence from the U.S.A. and U.S.S.R. *Canadian Journal of Psychology*, *37*, 565–571.
Boles, D.B. (1983). Dissociated imageability, concreteness, and familiarity in lateralized word recognition. *Memory and Cognition*, *11*, 511–519.
Boles, D.B. (1985). The effects of display and report order asymmetries on lateralized word recognition. *Brain and Language*, *26*, 106–116.
Boles, D.B., Rogers, S., & Wymer, W. (1982). Age of acquisition and visual field asymmetry in word recognition. *Perception and Psychophysics*, *32*, 486–490.
Bradley, D.C., & Garrett, M.F. (1983). Hemisphere differences in the recognition of closed and open class words. *Neuropsychologia*, *21*, 155–159.
Bradshaw, G.J., Hicks, R.E., & Rose, B. (1979). Lexical discrimination and letter-string identification in the two visual fields. *Brain and Language*, *8*, 10–18.
Bradshaw, J.L. (1980). Right-hemisphere language: A review of related issues. *Brain and Language*, *10*, 172–188.
Bradshaw, J.L., & Gates, E.A. (1978). Visual field differences in verbal tasks: Effects of task familiarity and sex of subject. *Brain and Language*, *5*, 166–187.
Bradshaw, J.L., Gates, A., & Nettleton, N.C. (1977). Bihemispheric involvement in lexical decisions: Handedness and a possible sex difference. *Neuropsychologia*, *15*, 277–286.
Bradshaw, J.L., Nettleton, N.C., & Taylor, M.J. (1981a). Right hemisphere language and cognitive deficit in sinistrals? *Neuropsychologia*, *19*, 113–132.
Bradshaw, J.L., Nettleton, N.C., & Taylor, M.J. (1981b). The use of laterally presented words in research into cerebral asymmetry: Is directional scanning likely to be a source of artifact? *Brain and Language*, *14*, 1–14.
Bradshaw, J.L., & Taylor, M.J. (1979). A word-naming deficit in non-familial sinistrals? Laterality effects of vocal responses to tachistoscopically presented letter strings. *Neuropsychologia*, *17*, 21–32.
Brand, N., Van Bekkum, I., Stumpel, M., & Kroeze, J.H.A. (1983). Word matching and lexical decisions: A visual half-field study. *Brain and Language*, *18*, 199–211.
Bryden, M.P., & Allard, F. (1976). Visual hemifield differences depend on typeface. *Brain and Language*, *3*, 191–200.
Bub, D.N. (1982). The nature of half-field asymmetry for word processing as a function of hemispheric specialization. Ph.D. dissertation, University of Rochester.
Burgess, C., & Simpson, G.B. (in press). Cerebral hemispheric mechanisms in the retrieval of ambiguous word meanings. *Brain and Language*.
Byrd, M., & Moscovitch, M. (1984). Lateralization of peripherally and centrally masked words in young and elderly people. *Journal of Gerontology*, *39*, 699–703.
Caplan, D., Holmes, J.M., & Marshall, J.C. (1974). Word classes and hemispheric

specialization. *Neuropsychologia*, *12*, 331–337.

Chambers, S.M., & Forster, K.I. (1975). Evidence for lexical access in a simultaneous matching task. *Memory and Cognition*, *3*, 549–559.

Chiarello, C. (1985). Hemisphere dynamics in lexical access: Automatic and controlled priming. *Brain and Language*, *26*, 146–172.

Chiarello, C. (in press). More on words, hemifields, and hemispheres: A reply to Schwartz & Kirsner. *Brain and Cognition*.

Chiarello, C., Church, K.L., & Hoyer, W.J. (1985). Automatic and controlled semantic priming: Accuracy, response bias, and aging *Journal of Gerontology*, *40*, 593–600.

Chiarello, C., Dronkers, N.F., & Hardyck, C. (1984). Choosing sides. On the variability of language lateralization in normal subjects. *Neuropsychologia*, *22*, 363–374.

Chiarello, C., & Nuding, S. (1987). Visual field effects for processing content and function words. *Neuropsychologia*, *25*, 539–548.

Chiarello, C., Nuding, S., & Pollock, A. (in press). Lexical decision and naming asymmetries: Influence of response selection and response bias. *Brain and Language*.

Chiarello, C., Senehi, J., & Nuding, S. (1987). Semantic priming with abstract and concrete words: Differential asymmetry may be postlexical. *Brain and Language*, *31*, 43–60.

Chiarello, C., Senehi, J., & Soulier, M. (1986). Viewing conditions and hemisphere asymmetry for the lexical decision. *Neuropsychologia*, *24*, 521–529.

Cohen, G., & Freeman, R. (1978). Individual differences in reading strategies in relation to cerebral asymmetries. In J. Requin (Ed.), *Attention and performance VII* (pp. 411–426). Hillsdale, NJ: Lawrence Erlbaum Associates.

Collins, A.M., & Loftus, E.F. (1975). A spreading-activation theory of semantic processing. *Psychological Review*, *82*, 407–428.

Coltheart, M. (1978). Lexical access in simple reading tasks in G. Underwood (Ed.), *Strategies of information processing* (pp. 151–216). London: Academic Press.

Coltheart, M. (1980a). Deep dyslexia: A right-hemisphere hypothesis in M. Coltheart, K. Patterson, & J.C. Marshall (Eds.), *Deep dyslexia* (pp. 326–380). London: Routledge & Kegan Paul.

Coltheart, M. (1980b). Reading, phonological recoding, and deep dyslexia. In M. Coltheart, K. Patterson, & J.C. Marshall (Eds.), *Deep dyslexia* (pp. 197–226). London: Routledge & Kegan Paul.

Coltheart, M., Davelaar, E., Jonasson, J.T., & Besner, D. (1977). Access to the internal lexicon. In S. Dornic (Ed.), *Attention and performance VI* (pp. 535–556). Hillsdale, NJ: Lawrence Erlbaum Associates.

Day, J. (1977). Right-hemisphere language processing in normal right-handers. *Journal of Experimental Psychology: Human Perception and Performance*, *3*, 518–528.

Day, J. (1979). Visual half-field word recognition as a function of syntactic class and imageability. *Neuropsychologia*, *17*, 515–520.

de Groot, A.M.B. (1984). Primed lexical decision: Combined effects of the proportion of related prime-target pairs and the stimulus-onset asynchrony of prime and target. *Quarterly Journal of Experimental Psychology*, *36A*, 253–280.

Dornbush, R.L., & Winnick, W.A. (1965). Right-left differences in tachistoscopic identification of paralogs as a function of order of approximation to English letter sequences. *Perceptual and Motor Skills*, *20*, 1222–1224.

Eglin, M. (1987). Interference and priming within and across visual fields in a lexical decision task. *Neuropsychologia*, *25*, 613–623.

Ellis, H.D., & Shepherd, J.W. (1974). Recognition of abstract and concrete words presented in left and right visual fields. *Journal of Experimental Psychology*, *103*, 1035–1036.

Ellis, H.D., & Young, A.W. (1977). Age-of-acquistion and recognition of nouns presented in the left and right visual fields: A failed hypothesis. *Neuropsychologia*, *15*, 825–828.

Elman, J.L., Takahashi, K., & Tohsaku, Y. (1981). Lateral asymmetries for the identification of concrete and abstract Kanji. *Neuropsychologia*, *19*, 407–412.

Evett, L.J., & Humphreys, G.W. (1981). The use of abstract graphemic information in lexical access. *Quarterly Journal of Experimental Psychology*, *33a*, 325–350.

Fischler, I. (1977). Associative facilitation without expectancy in a lexical decision task. *Journal of Experimental Psychology: Human Perception and Performance*, *3*, 18–26.

Fischler, I., & Goodman, G.O. (1978). Latency of associative facilitation in memory. *Journal of Experimental Psychology: Human Perception and Performance*, *4*, 455–470.

Forster, K.I. (1976). Accessing the internal lexicon. In R.J. Wales and E. Walker (Eds.), *New approaches and language mechanisms* (pp. 256–287). Amsterdam: New Holland.

Forster, K.I. (1978). Accessing the mental lexicon. In E. Walker (Ed.), *Explorations in the biology of language* (pp. 139–174). Montgomery, VT: Bradford Books.

Fowler, C.A., Wolford, G., Slade, R., & Tassinary, L. (1981). Lexical access with and without awareness. *Journal of Experimental Psychology: General*, *110*, 341–362.

Frederiksen, J.R., & Kroll, J.F. (1976). Spelling and sound: Approaches to the internal lexicon. *Journal of Experimental Psychology: Human Perception and Performance*, *2*, 361–379.

Gazzaniga, M.S. (1983a). Right hemisphere language following brain bisection. *American Psychologist*, *38*, 525–537.

Gazzaniga, M.S. (1983b). Reply to Levy and to Zaidel. *American Psychologist*, *38*, 547–549.

Gernsbacher, M.A. (1984). Resolving 20 years of inconsistent interactions between lexical familiarity and orthography, concreteness, and polysemy. *Journal of Experimental Psychology: General*, *113*, 256–281.

Gibson, A.R., Dimond, S.J., & Gazzaniga, M.S. (1972). A left field superiority for word matching. *Neuropsychologia*, *10*, 463–466.

Gilhooly, K.J., & Gilhooly, M.L. (1979). Age-of-acquisition effects in lexical and episodic memory tasks. *Memory and Cognition*, *7*, 214–223.

Gill, K.M., & McKeever, W.F. (1974). Word length and exposure time effects on the recognition of bilaterally presented words. *Bulletin of the Psychonomic Society*, *4*, 173–175.

Glushko, R.J. (1979). The organization and activation of orthographic knowledge of reading aloud. *Journal of Experimental Psychology: Human Perception and Performance*, *5*, 674–691.
Goodall, G. (1984). Morphological complexity and cerebral lateralization. *Neuropsychologia*, *22*, 375–380.
Goodglass, H., & Menn, L. (1985). Is agrammatism a unitary phenomenon? In M. Kean (Ed.), *Agrammatism* (pp. 1–26). Orlando FL: Academic Press.
Gordon, B. (1983). Lexical access and lexical decision: Mechanisms of frequency sensitivity. *Journal of Verbal Learning and Verbal Behavior*, *22*, 24–44.
Gordon, B., & Caramazza, A. (1982). Lexical decision for open- and closed-class words: Failure to replicate differential frequency sensitivity. *Brain and Language*, *15*, 143–160.
Gordon, B., & Caramazza, A. (1985). Lexical access and frequency sensitivity: Frequency saturation and open/closed class equivalence. *Cognition*, *21*, 95–115.
Gross, M.M. (1972). Hemispheric specialization for processing of visually presented verbal and spatial information. *Perception and Psychophysics*, *12*, 357–363.
Hardyck, C., Chiarello, C., Dronkers, N.F., & Simpson, G.V. (1985a). Orienting attention within visual fields: How efficient is interhemispheric transfer? *Journal of Experimental Psychology: Human Perception and Performance*, *11*, 650–666.
Hardyck, C., Dronkers, N.F., Chiarello, C., & Simpson, G.V. (1985b). The eyes have it: Exposure times and saccadic movements in visual half-field experiments. *Brain and Cognition*, *4*, 430–438.
Hatta, T. (1977a). Lateral recognition of abstract and concrete Kanji in Japanese. *Perceptual and Motor Skills*, *45*, 731–734.
Hatta, T. (1977b). Recognition of Japanese Kanji in the left and right visual fields. *Neuropsychologia*, *15*, 685–688.
Healey, J.M., Waldstein, S., & Goodglass, H. (1985). Sex differences in the lateralization of language discrimination vs. language production. *Neuropsychologia*, *23*, 777–790.
Heister, G. (1983). Visual half-field superiority for righthanders with familial sinistrality: Sex differences and cognitive/motor interference in a lexical decision task. *International Journal of Neuroscience*, *19*, 293–308.
Heister, G. (1984). Sex differences and cognitive/motor interference with visual half-field stimulation. *Neuropsychologia*, *22*, 205–214.
Heister, G., Kolbert, C., & Hofmeister, K. (1983). Sex differences and asymmetry of lexical processing: Effects of responding hand, stimulus familiarity and intra-experimental experience. *International Journal of Neuroscience*, *21*, 1–14.
Henderson, L. (1982). *Orthography and word recognition in reading*. London: Academic Press.
Hillinger, M.L. (1980). Priming effects with phonemically similar words: The encoding-bias hypothesis reconsidered. *Memory and Cognition*, *8*, 115–123.
Hines, D. (1976). Recognition of verbs, abstract nouns and concrete nouns from the left and right visual half-fields. *Neuropsychologia*, *14*, 211–216.
Hines, D. (1977). Differences in tachistoscopic recognition between abstract and concrete words as a function of visual half-field and frequency. *Cortex*, *13*, 66–73.
Hines, D. (1978). Visual information processing in the left and right hemispheres. *Neuropsychologia*, *16*, 593–600.

Hudson, P.T.W., & Bergman, M.W. (1985). Lexical knowledge in word recognition: Word length and word frequency in naming and lexical decision times. *Journal of Memory and Language*, *24*, 46–58.

James, C.T. (1975). The role of semantic information in lexical decisions. *Journal of Experimental Psychology: Human Perception and Performance*, *104*, 130–136.

Klein, D., Moscovitch, M., & Vigna, C. (1976). Attentional mechanisms and perceptual asymmetries in tachistoscopic recognition of words and faces. *Neuropsychologia*, *14*, 55–66.

Klein, R.M., & Smith, L.C. (1985). Is the emergence of a right visual field advantage in the category matching task dependent upon category constancy? *Canadian Journal of Psychology*, *39*, 88–99.

Koriat, A. (1984). Reading without vowels: Lexical access in Hebrew. In H. Bouma and D.G. Bouwhuis (Eds.), *Attention and performance X. Control of language processes* (pp. 227–242). Hillsdale, NJ: Lawrence Erlbaum Associates.

Kreuger, L.E. (1975). The word-superiority effect: Is its locus visual-spatial or verbal? *Bulletin of the Psychonomic Society*, *6*, 465–468.

Kroll, J.F., & Merves, J.S. (1986). Lexical access for concrete and abstract words. *Journal of Experimental Psychology: Learning, Memory, and Cognition*, *12*, 92–107.

Lambert, A.J., & Beaumont, J.G. (1983). Imageability does not interact with visual field in lateral word recognition with oral report. *Brain and Language*, *20*, 115–142.

Leiber, L. (1976). Lexical decisions in the right and left cerebral hemispheres. *Brain and Language*, *3*, 443–450.

Levy, J. (1983). Language, cognition, and the right hemisphere. *American Psychologist*, *38*, 538–541.

Mackavey, W., Curcio, F., & Rosen, J. (1975). Tachistoscopic word recognition performance under conditions of simultaneous bilateral presentation. *Neuropsychologia*, *13*, 27–33.

Marcel, A.J., & Patterson, K.E. (1978). Word recognition and production: Reciprocity in clinical and normal studies. In J. Requin (Ed.), *Attention and performance VII* (pp. 209–226). Hillsdale, NJ: Lawrence Erlbaum Associates.

McCusker, L.X., Hillinger, M.L., & Bias, R.G. (1982). Phonological recoding and reading. *Psychological Bulletin*, *89*, 217–245.

McKeever, W.F. (1971). Lateral word recognition: Effects of unilateral and bilateral presentation, asynchrony of bilateral presentation, and forced order of report. *Quarterly Journal of Experimental Psychology*, *23*, 410–416.

McKeever, W.F., & Gill, K.M. (1972). Visual half-field differences in the recognition of bilaterally presented single letters and vertically spelled words. *Perceptual and Motor Skills*, *34*, 815–818.

McKeever, W.F., & Hoff, A.L. (1982). Familial sinistrality, sex, and laterality differences in naming and lexical decision latencies of right-handers. *Brain and Language*, *17*, 225–239.

Meyer, D.E., & Schvaneveldt, R.W. (1971). Facilitation in recognizing pairs of words: Evidence of a dependence between retrieval operations. *Journal of Experimental Psychology*, *90*, 227–234.

Michaels, C.F., & Turvey, M.T. (1979). Central sources of visual masking: Indexing structures supporting seeing of a single, brief glance. *Psychological Research*, *41*, 1–61.

Morton, J. (1969). Interaction of information in word recognition. *Psychological Review*, *76*, 165–178.

Morton, J., & Patterson, K. (1980). A new attempt at an interpretation or, an attempt at a new interpretation. In M. Coltheart, K. Patterson, & J.C. Marshall (Eds.), *Deep dyslexia* (pp. 91–118). London: Routledge & Kegan Paul.

Moscovitch, M. (1981). Right hemisphere language. *Topics in Language Disorders*, *1*, 41–62.

Moscovitch, M. (1983). Laterality and visual masking: Interhemispheric communication and the locus of perceptual asymmetries for words. *Canadian Journal of Psychology*, *37*, 85–106.

Myers, J.J. (1984). Right hemisphere language: Science or fiction? *American Psychologist*, *39*, 315–320.

Neely, J.H. (1977). Semantic priming and retrieval from lexical memory: Roles of inhibitionless spreading activation and limited-capacity attention. *Journal of Experimental Psychology, General*, *106*, 226–254.

Orenstein, H.B., & Meighan, W.B. (1976). Recognition of bilaterally presented words varying in concreteness and frequency: Lateral dominance or sequential processing? *Bulletin of the Psychonomic Society*, *7*, 179–180.

Oscar-Berman, M., Goodglass, H., & Cherlow, D.G. (1973). Perceptual laterality and iconic recognition of visual materials by Korsakoff patients and normal adults. *Journal of Comparative and Physiological Psychology*, *82*, 316–321.

Paivio, A. (1971). *Imagery and verbal processes*. New York: Holt, Rinehart, & Winston.

Paivio, A., & O'Neill, B.J. (1970). Visual recognition thresholds and dimensions of word meaning. *Perception and Psychophysics*, *8*, 273–275.

Paivio, A., & te Linde, J. (1982). Imagery, memory, and the brain. *Canadian Journal of Psychology*, *36*, 243–272.

Parkin, A.J. (1984). Redefining the regularity effect. *Memory and Cognition*, *12*, 287–292.

Parkin, A.J., & West, S. (1985). Effects of spelling-to-sound regularity on word identification following brief presentation in right or left visual field. *Neuropsychologia*, *23*, 270–284.

Patterson, K., & Besner, (1984a). Is the right hemisphere literate? *Cognitive Neuropsychology*, *1*, 315–342.

Patterson, K., & Besner, D. (1984b). Reading from the left: A reply to Rabinowicz and Moscovitch and Zaidel and Schweiger. *Cognitive Neuropsychology*, *1*, 365–380.

Patterson, K.E., & Kay, J. (1982). Letter-by-letter reading: Psychological descriptions of a neurological syndrome. *Quarterly Journal of Experimental Psychology*, *34A*, 411–441.

Pring, T.R. (1981). The effect of stimulus size and exposure duration on visual field asymmetries. *Cortex*, *17*, 227–240.

Purcell, D.G., & Stanovitch, K.E. (1982). Some boundary conditions for a word superiority effect. *Quarterly Journal of Experimental Psychology*, *34A*, 117–134.

Rabinowicz, B., & Moscovitch, M. (1984). Right hemisphere literacy: A critique of some recent approaches. *Cognitive Neuropsychology*, *1*, 343–350.

Regard, M., Landis, T., & Graves, R. (1985). Dissociated hemispheric superiorities for reading stenography vs. print. *Neuropsychologia*, *23*, 431–435.

Richardson, J.T.E. (1976). The effects of stimulus attributes upon latency of word recognition. *British Journal of Psychology*, *67*, 315–325.

Riegel, K.F., & Riegel, R.M. (1961). Prediction of word-recognition thresholds on the basis of stimulus-parameters. *Language and Speech*, *4*, 157–170.

Rodel, M., Dudley, J.G., & Bourdeau, M. (1983). Hemispheric differences for semantically and phonologically primed nouns: A tachistoscopic study in normals. *Perception and Psychophysics*, *34*, 523–531.

Rosson, M.B. (1985). The interaction of pronunciation rules and lexical representations in reading aloud. *Memory and Cognition*, *13*, 90–98.

Saffran, E.M., Bogyo, L.C., Schwartz, M.F., & Marin, O.S.M. (1980). Does deep dyslexia reflect right-hemisphere reading? In M. Coltheart, K. Patterson, & J.C. Marshall (Eds.), *Deep dyslexia* (pp. 381–406). London: Routledge & Kegan Paul.

Sasanuma, S., Itoh, M., Kobayashi, Y., & Mori, K. (1980). The nature of task-stimulus interaction in the tachistoscopic recognition of Kana and Kanji words. *Brain and Language*, *9*, 298–308.

Sasanuma, S., Itoh, M., Mori, K., & Kobayashi, Y. (1977). Tachistoscopic recognition of Kana and Kanji words. *Neuropsychologia*, *15*, 547–553.

Schiepers, C. (1980). Response latency and accuracy in visual word recognition. *Perception and Psychophysics*, *27*, 71–81.

Schmuller, J., & Goodman, R. (1979). Bilateral tachistoscopic perception, handedness and laterality. *Brain and Language*, *8*, 81–91.

Seidenberg, M.S. (1985). The time course of information activation and utilization in visual word recognition. In D. Besner, T.G. Waller, & G.E. Mackinnon (Eds.), *Reading Research: Advances in Theory and Practice, Vol. 5*, (pp. 199–252). Orlando FL: Academic Press.

Seidenberg, M.S., Waters, G.S., Barnes, M.A., & Tanenhaus, M.K. (1984). When does irregular spelling or pronunciation influence word recognition? *Journal of Verbal Learning and Verbal Behavior*, *23*, 383–404.

Seidenberg, M.S., Waters, G.S., Sanders, M., & Langer, P. (1984). Pre- and postlexical loci of contextual effects on word recognition. *Memory and Cognition*, *12*, 315–328.

Sergent, J. (1983a). The effects of sensory limitations on hemispheric processing. *Canadian Journal of Psychology*, *37*, 345–366.

Sergent, J. (1983b). The role of the input in visual hemispheric asymmetries. *Psychological Bulletin*, *93*, 481–512.

Sergent, J. (1984). Role of contrast, lettercase, and viewing conditions in a lateralized word-naming task. *Perception and Psychophysics*, *35*, 489–498.

Shanon, B. (1979a). Lateralization effects in lexical decision tasks. *Brain and Language*, *8*, 380–387.

Shanon, B. (1979b). Lateralization effects in response to words and nonwords. *Cortex*, *15*, 541–549.

Sperry, R.W., Gazzaniga, M.S., & Bogen, J.E. (1969). Interhemispheric relationships: The neocortical commisures, syndromes of hemispheric disconnection. In P.J. Vinken & G.W. Bruyn (Eds.), *Handbook of clinical neurology*, *Vol. 4* (pp. 273–290). Amsterdam: Elsevier/North Holland.

Stanners, R.F., & Forbach, G.B. (1973). Analysis of letter strings in word recognition. *Journal of Experimental Psychology*, *98*, 31–35.

Tomlinson-Keasey, C., Brewer, A., & Huffman, K. (1983). The importance of being first: An analysis of tachistoscopic presentations of words. *Cortex*, *19*, 309–326.

Turvey, M.T. (1973). On peripheral and central processes in vision: Inferences from an information processing analysis of masking with patterned stimuli. *Psychological Review*, *80*, 1–52.

Urcuioli, P., Klein, R., & Day, J. (1981). Hemispheric differences in semantic processing: Category matching is not the same as category membership. *Perception & Psychophysics*, *29*, 343–351.

Walker, E., & Ceci, S.J. (1985). Semantic priming effects for stimuli presented to the right and left visual fields. *Brain and Language*, *25*, 144–159.

Winnick, W.A., & Kressel, K. (1965). Tachistoscopic recognition thresholds, paired-associate learning, and free recall as a function of abstractness-concreteness and word frequency. *Journal of Experimental Psychology*, *70*, 163–168.

Young, A.W., & Bion, P.J. (1980). Hemifield differences for naming bilaterally presented nouns varying on age of acquisition. *Perceptual and Motor Skills*, *50*, 366.

Young, A.W., Bion, P.J., & Ellis, A.W. (1980). Studies toward a model of laterality effects for picture and word matching. *Brain and Language*, *11*, 54–65.

Young, A.W., Bion, P.J., & Ellis, A.W. (1982). Age of reading acquisition does not affect visual hemifield asymmetries for naming imageable words. *Cortex*, *18*, 477–483.

Young, A.W., & Ellis, A.W. (1985). Different methods of lexical access for words presented in the left and right visual hemifields. *Brain and Language*, *24*, 326–358.

Young, A.W., Ellis, A.W., & Bion, P.J. (1984). Left hemisphere superiority for pronounceable nonwords but not unpronounceable nonwords. *Brain and Language*, *22*, 14–25.

Zaidel, E. (1983a). Disconnection syndrome as a model for laterality effects in the normal brain. In J.B. Hellige (Ed.), *Cerebral hemisphere asymmetry: Methods, theory and application* (pp. 95–151). New York: Praeger.

Zaidel, E. (1983b). A response to Gazzaniga. *American Psychologist*, *38*, 542–546.

Zaidel, E., & Schweiger, A. (1984). On wrong hypotheses about the right hemisphere: Commentary on K. Patterson & D. Basner, "Is the right hemisphere literate?" *Cognitive Neuropsychology*, *1*, 351–364.

3
Neuropsychology of Spatial Memory

WILLIAM W. BEATTY and ALEXANDER I. TRÖSTER

In recent years the study of spatial memory has occupied the attention of many cognitive scientists from a wide range of disciplines. For some, the study of spatial memory is an end in its own right. For others, measures of spatial learning and memory provide a convenient index of the development and decline of intellectual functions and their neurobiological correlates throughout the life span. In the present review we attempt to summarize current knowledge of the neural mechanisms underlying spatial behavior, drawing on information from studies of nonhuman species as well as humans. Such an endeavor is beset by all the usual uncertainties regarding comparability of behavioral tests and neuroanatomical homologies. In addition, the comparative neuropsychology of spatial behavior is complicated by the rather different perspectives of space imposed on animals of vastly different size. The view of the world of rats and humans is obviously quite different. In an attempt to minimize confusion resulting from differing motivational conditions we arbitrarily excluded studies in which escape from or avoidance of electric shock served as reinforcement because aversive conditions of this sort have rarely been used in species other than rats.

Although many ways of classifying spatial behavior have been proposed, probably the most widely accepted is the distinction between egocentric and allocentric frames of reference. An egocentric reference system is one in which the origin of the system is the person's own body and the location of other points in space is known only with respect to one's self. An allocentric frame of reference is one in which the relationship of points in the environment with respect to each other is known independent of the position of the observer. As this seems to be a fundamental distinction, it is reasonable to hypothesize that two separable neural systems, one subserving egocentric spatial relations, the other allocentric spatial relations, might be identifiable. Indeed, models have been proposed assigning egocentric functions to one region of the brain and allocentric functions to another. Though this simple scheme is clearly not supported by the data, the possibility of separate but closely interrelated systems for egocentric and allocentric functions remains a viable theoretical option. A useful manner to examine the neural basis of spatial memory is to examine the effects on spatial behavior of lesions in different brain regions, as well as of pharmacological agents and aging.

Temporal Lobe

The importance of the temporal lobe for memory in general and spatial memory in particular is well established. In 1953 patient H.M. underwent bilateral resection of the medial temporal lobes for the relief of intractable epilepsy. The exact extent of the brain damage was not reported but it is presumed to include the prepyriform gyrus, uncus, amygdala, and portions of the hippocampus and parahippocampal gyrus (Scoville & Milner, 1957.) After the operation, H.M. developed severe anterograde amnesia for both verbal and nonverbal material. For example, he was unable to learn the correct path from start to finish on a stylus maze, regardless of whether presentation was in a visual (Milner, 1965) or tactual format (Corkin, 1965). Likewise, H.M.'s memory for the locations of objects in visual space is severely impaired (Smith & Milner, 1981). Because H.M. also exhibits some deficits on visual–spatial tasks that impose a mild memory load such as the hidden figures test and copying the Rey-Osterreith figure, it is difficult to determine the extent that his perceptual deficits contribute to his spatial memory problems. On the other hand, he performs normally on perceptual tasks such as the Mooney faces test and the block design test on the Wechsler-Bellvue. Furthermore, he was able to draw an accurate floor plan of the house he lived in from 1960 to 1974 (Corkin, 1984). These observations suggest that perceptual deficits make only a minor contribution to H.M.'s spatial memory defect. Further, he is clearly capable of learning and remembering some new visuospatial information, such as the floor plan of his house, albeit with great difficulty.

Considerable attention has been devoted to comparing the effects of unilateral damage of the left and right temporal lobes on learning and memory for verbal and spatial information. The patients in these experiments had received resections of varying amounts of the left or right temporal lobe for the relief of focal epilepsy. These studies generally support the conclusion that excision of the right temporal lobe impairs spatial learning and memory while comparable damage to the left temporal lobe has almost no effect. Thus, right temporal lobectomy impairs learning of visual or tactual stylus mazes (Corkin, 1965; Milner, 1965) and memory for visual locations (Milner, 1974). On the latter task, the right temporal lobe group performed normally when spatial memory was tested after a short delay (6 seconds), but became increasingly impaired as the retention interval was lengthened. This demonstrates that visuoperceptual deficits, which can occur in right temporal lobectomy patients (e.g., Lansdell,

1968), cannot explain the memory deficit. Furthermore, the right temporal lobectomy patients in this study performed normally on a verbal short-term memory task on which left temporal lobectomy patients were impaired, so the deficits of both groups were clearly not global. Finally, there is a clear relationship between the extent of right temporal lobe damage and severity of the spatial memory defect. Patients whose excisions involved minimal damage to the right hippocampus were not impaired while patients with extensive damage to the hippocampus displayed serious disturbances of spatial memory. Petrides (1985) found that patients with left temporal lobe lesions were impaired in learning nonspatial conditional tasks, and patients with right temporal lobe lesion were impaired on a spatial conditional response task (touching one of six cards in response to one of six lights placed in an irregular array). In both instances, extensive hippocampal involvement was required for deficits to be apparent. By contrast, frontal lobectomy patients, irrespective of lesion lateralization, were impaired on both the spatial and nonspatial conditional tasks. In cases of involvement of the posterior dorsolateral frontal cortex, impairments were more marked. It is of note that frontal lobectomy patients' deficits cannot be explained in terms of lack of initiative, poor attention, or not following instructions, as none of the patients had difficulty in discriminating between the stimuli, or in making the movements required for responses during testing despite showing impairment in learning to select the correct movement to a given stimulus. In the spatial task, subjects who had first been given the nonspatial task, did show some advantage but were still severely impaired.

Normal subjects have been found to encode spatial locations relatively automatically (Mandler, Seegmiller, & Day, 1977), although this has recently been called into question by Cooper and Marshall (1985). Using a task in which the location of objects on a two-dimensional array was learned only incidentally, Smith and Milner (1981) found that right temporal lobe lesions impaired memory for spatial location but left temporal lobe lesions did not have this effect. Again deficits were observed only if the lesions involved radical excision of the hippocampus.

Thus, in humans the data indicate that the integrity of the medial temporal lobe of the right hemisphere to relatively more important for spatial than for verbal memory while the left temporal lobe is more involved with verbal than with spatial memories. These specializations are relative, and may be restricted to tasks in which stimuli are presented visually. Auditory verbal memory is about equally affected by right and left temporal lobe damage (Samuels, Butters, & Fedio, 1972; Samuels, Butters, Fedio, & Cox, 1980), but right temporal lobectomy impairs memory for designs (Jones-Gotman, 1986a; 1986b). Furthermore, patient H.M. whose temporal lobe damage is bilateral, performs consistently below the worst of the right temporal lobectomy patients on all tests of spatial memory. It is likely that damage to the hippocampal formation is responsible for the deficits in spatial memory that accompany temporal lobe damage in humans, but this conclusion cannot be drawn with certainty because all of the patients with hippocampal damage also sustained damage to the amygdala.

Studies with patients with temporal lobe damage have focused on their deficits in anterograde spatial memory, primarily on tasks that require remembering allocentric information. Major deficits in personal or extrapersonal orientation following temporal lobe damage seem unlikely in as much as Butters, Soeldner, and Fedio (1972) found that patients with unilateral temporal lobectomies performed normally on the Money Road Map test (a task that measures egocentric orientation) and the stick reversal task (Butters & Barton, 1970), which measures extrapersonal orientation. On the other hand H.M. exhibits deficits on some tests of both egocentric and allocentric orientation, but performs normally on others (Corkin, 1984).

To our knowledge, remote memory for spatial information has been studied in only one human patient with temporal lobe injury. The patient received a radical right-sided temporal lobectomy to remove a massive glioblastoma. Postoperatively she displayed marked visuoperceptual difficulties and anterograde memory impairments that were more severe for nonverbal than for verbal tasks. Geographical knowledge, both verbal and visuospatial, was unimpaired. (Beatty, MacInnes, Porphyris, Tröster, & Cermak in press). This raises the possibility that the right temporal lobe is cruicial to the formation of anterograde spatial memories, but not for the retrieval of such memories once they have been laid down. This issue has been the subject of heated debate in the animal literature which is reviewed next.

Animal Studies

A number of different tests have been used to study spatial learning and memory in animals. For nonhuman primates and carnivores, the animals are typically caged in front of a tray or panel that contains the test apparatus. This arrangement is roughly comparable to the arrangement of most testing situations with humans. A somewhat analogous setting for rats can be created in an operant conditioning chamber by placing discriminative stimuli and response manipulanda on a panel in front of the animal. In most of these testing arrangements the subject can, at least potentially, use its own body as a reference point to maintain appropriate orientation, restricting testing to egocentric spatial abilities.

A great deal of the research on spatial behavior in rats has employed mazes of varying complexity. Because the animal must move through the maze, the relationship between the location of its body and the rest of the experimental space is constantly changing, permitting an assessment of allocentric spatial memory.

With few exceptions, animals with lesions of either the amygdala or hippocampus acquire simultaneous spatial discriminations normally (Mahut, 1971; cf. Gray & McNaughton, 1983; O'Keefe & Nadel, 1978 for other references). Deficits may be observed in animals with combined lesions of the amygdala and either the hippocampus or the anterior temporal cortex (Mahut, 1971; Jones & Mishkin, 1972).

By contrast, spatial reversal learning is generally impaired and lesions restricted to the hippocampus are sufficient to produce the impairment. Generally comparable results have been reported in rats and monkeys (Mahut, 1971; Samuels, 1972). Furthermore lesions of the septal area (Bengelloun, Nelson, Gerth, & Beatty, 1975) or transections of the fornix (Mahut & Zola, 1973) produce a similar pattern of impaired spatial reversal learning but normal acquisition. In the spatial reversal task there is a consistent relationship between the correct choice and bodily orientation so the implication is that bilateral hippocampal damage can impair this aspect of egocentric spatial behavior.

The spatial delayed response task has been widely employed in studies with primates. Early research on temporal lobe lesions produced equivocal results. For example, Mishkin (1954) observed no clear-out impairment following hippocampal lesions in monkeys that had learned the task preoperatively. Similarly no impairments were reported in monkeys with medial temporal lobe ablations (combined lesions of amygdala and hippocampus) that were trained postoperatively (Mahut & Cordeau, 1963; Orbach, Milner, & Rasmussen, 1960). Recently Zola-Morgan and Squire (1985a) obtained clear evidence of impairment on delayed response performance in monkeys with combined amygdalohippocampal lesions that were trained postoperatively. Their data demonstrated that the length of the delay interval was a critical variable. The deficient performance by the animals with medial temporal lobe lesions was observed only with delays of 15 seconds or longer. They suggested that previous failures to observe deficits after medial temporal lobe lesions could be attributed to the use of short delays. It is not yet clear whether or not lesions that are restricted to the amygdala or the hippocampus will impair performance on the delayed response task. Nor is it clear if animals trained preoperatively would experience similar deficits if the delay were sufficiently long.

The effects of temporal lobe lesions on delayed alternation are somewhat more consistent. In monkeys, deficits have been observed following large medial temporal lobe lesions (Mahut & Cordeau, 1963; Orbach et al., 1960) as well as with smaller lesions restricted to the amygdala (Mahut, 1971) or hippocampus (Rosvold & Szwarcbart, 1964). However, not all monkeys with hippocampal lesions are impaired on delayed alternation (Waxler & Rosvold, 1970). Apparently some monkeys adopted the strategy of pressing one lever and then immediately shifting to the opposite lever to await the next trial.

When delayed alternation has been studied in mazes, rats with hippocampal lesions are almost always impaired (Gray & McNaughton, 1983), although occasionally the deficit is only apparent when the animal has to engage in some sort of distracting activity during the retention interval (Jarrard, 1975). Interpretation of this finding is complicated because considerable evidence indicates that rats with hippocampal lesions are less likely to alternate their choices spontaneously in a T maze or some similar apparatus (cf. O'Keefe & Nadel, 1978). As with spatial reversal learning, lesions of the fornix (Gaffan, 1972), the septal area (Thomas & Spafford, 1984) or the mammillothalamic tracts (Thomas & Gash, 1985) have much the same effect.

In the operant chamber, the effects of hippocampal lesions in rats on delayed alternations are quite variable. Both deficits and facilitation have been reported (Niki, 1966; Stevens & Cowey, 1972). These confusing results are difficult to interpret because control animals often did not attain performance levels above 70%, suggesting that they had not learned the alternation rule. A further complication is that the ages of animals are not always stated and the production of deficits after hippocampal lesioning may be age-dependent. For example, delayed response and reversal learning deficits characteristic of hippocampal lesions in adult rats are not evident when the same lesions are made in infants (Isaacson, Nonneman, & Schmaltz, 1968). Overall, the results of delayed response and delayed alteration experiments support the view that animals with temporal lobe damage can maintain egocentric orientation adequately, but experience difficulty when a memory component is added to the task.

Much of the recent research on spatial memory in rats has employed complex mazes that afford better opportunity for designing analytical experiments to understand the nature of lesion-induced deficits. One such apparatus is the elevated multiple-arm radial maze. This apparatus resembles a rimless wagon wheel set on a pedestal in the horizontal plane and is usually housed in a room with many extramaze cues. In the most common testing procedure each identical arm is baited with food at the outset of the test session. As food is never replaced during the session, to attain optimal performance the animal must adopt a "win-shift" rule, never reentering arms already visited. Normal adult rats require only 10–15 sessions to reach a criterion of consistently entering seven different arms on their first eight choices. Accurate performance apparently depends on memory for the relationship between arms visited and visual cues in the extramaze environment. In support of this conclusion are the results of an extensive series of control experiments that indicate that neither intramaze stimuli nor response strategies are essential for accurate performance. To break up response sequences, rats have been confined to the center of the maze between choices (Olton, Collison, & Werz, 1977), removed from the maze between choices (Maki, Brokofsky, & Berg, 1979), or forced to visit a randomly chosen subset of arms at the outset of the session (Beatty & Shavalia, 1980). None of these manipulations affected performance. The importance of extra- as opposed to intramaze cues has been demonstrated by experiments in which the animals are allowed to make several choices and then the maze is rotated to place chosen arms in unchosen locations and unchosen arms in chosen locations. In such experiments rats avoid previously visited locations and enter previously visited arms, indicating the importance of extramaze cues for accurate performance. (Maki, Beatty, Hoffman, Bierley, & Clouse, 1984; Olton & Collison, 1979). Thus their choices are guided by spatial as opposed to nonspatial cues.

The basic task can be modified to permit a simultaneous test of both working and reference memory. To accomplish, this only a subset of arms is ever baited with food. The same arms are unbaited from day to day so the rat must learn never to visit the unbaited arms. Because this is a stable rule, entries into unbaited arms constitute reference memory errors. Because rats do not visit baited arms in any consistent pattern (or can be forced not to do so), there is no rule that can be applied from day to day. The rat must remember which of the baited arms it has already visited. Reentries into baited arms constitute working memory errors.

Another task widely used to study spatial reference memory is the Morris water maze (Morris, 1981). The apparatus is a circular pool filled with cool opaque fluid housed in a room equipped with a rich variety of extramaze cues. A small platform is submerged beneath the surface. When first placed into the water maze, normal rats systematically swim around the maze until they find the hidden platform. After 4–8 trials rats swim directly to the platform regardless of the place from which they enter the pool. This sort of performance could only occur if the rat could remember the spatial relationships between the platform and the extramaze stimuli (allocentric spatial information). Hence the Morris water maze provides a relatively pure test of spatial mapping ability.

Using the basic procedure for testing only working memory in the radial maze, Olton, Walker, and Gage (1978) demonstrated that lesions of the fimbria–fornix, septum, or entorhinal cortex produced profound loss of accuracy of spatial working memory in rats that were trained preoperatively. No recovery was observed over 50 days of postoperative testing. In subsequent work, Becker, Walker, and Olton (1980) showed that neither amygdaloid nor caudate lesions affected performance. These findings indicated that the integrity of neural circuits through the hippocampal formation is critical to accurate spatial working memory. Olton, Walker, and Wolf (1982) confirmed this idea in a disconnection experiment that revealed that the integrity of a single ipsilateral or contralateral pathway through the hippocampal formation was sufficient to permit accurate spatial memory.

Attempts to determine which cell fields within the hippocampus are critical to spatial memory in the radial maze have not met with great success. Jarrard (1978) found that while lesions of field CA1 or of the alveus (which contains efferents to the subiculum) impaired postoperative acquisition on the radial maze, such lesions did not affect performance of animals that acquired accurate performance preoperatively. Only complete hippocampal ablations or lesions of the fimbria disrupted preoperatively acquired behavior. This indicates that integrity of the CA1 field may only be required for encoding, but not storage or retrieval of spatial information.

Some support for the idea that normal hippocampal functioning is especially important for the encoding of spatial information comes from the work of Bierley, Kesner, and Novak (1983). These investigators observed disruption of accurate performance in rats trained on a spatial delayed matching to sample task following subseizure levels of electrical stimulation of the dorsal hippocampus. The deficit was observed only if the hippocampus was stimulated during the presentation of the sample and only if a 20-minute retention interval was employed. With a shorter delay (1–12 minutes) the stimulation was ineffective. Hippocampal stimulation after the sample was presented was ineffective at all of the retention intervals examined. On the other hand, stimulation of the hippocampus that is sufficiently intense to produce local seizure activity (without observable motor seizures) can produce retrograde amnesia when administered after the initial choices have been completed in an 8- or a 12-arm radial maze (Knowlton, McGowan, Olton, & Gamzu, 1985; Olton & Wolf, 1981). Because spatial memory was equally disrupted by stimulation at various times within an 8-hour-long retention interval, the deficit presumably reflects impairment of storage, retrieval, or both. Because behavioral deficits did not results from subseizure stimulation, it is impossible to conclude that hippocampal function was selectively affected. However, seizure-producing stimulation of the amygdala had no behavioral consequences (Olton & Wolf, 1981), implying that abnormal activity in the hippocampus or its major targets was responsible for the behavioral impairment.

Hippocampal Function in Spatial Behavior: Working Memory or Cognitive Mapping?

Since 1979 much research has focused on evaluating two major hypotheses concerning hippocampal function. The cognitive mapping theory of O'Keefe and Nadel (1978) postulates that the hippocampus is the center of the "locale" system that is essential to the absolute representation of spatial relationships. The corollary to this theory states that without the hippocampus the organism would be incapable of true place learning, but cue or response learning would remain intact. The discovery of "place" cells within the CA1 and CA3 fields, which fired when the organism was in a particular part of the environment (O'Keefe, 1976; Ranck, 1973), provided impressive neurophysiological evidence for the cognitive map hypothesis. From the perspective of cognitive mapping theory, deficits in working memory on the radial maze after damage to the hippocampus occur because the rat cannot perform the truly allocentric spatial analysis required by the task. Alternatively, Olton, Becker, and Handelman (1979) argued that the hippocampus is essential for accurate working memory. According to this hypothesis, rats with hippocampal lesions fail the radial maze problem not because the task requires spatial mapping, but because it requires working memory.

In principle, deciding between the two theories is quite simple. If the working memory hypothesis is correct, then animals with hippocampal lesions should be unimpaired on tasks that require spatial mapping but place no demand on working memory. Further, damage to the hippocampal system should impair performance on nonspatial tasks that require working memory. Cognitive mapping theory predicts deficits on all tasks that require true place learning and only on such tasks.

As a test of nonspatial working memory, Olton and Feustle (1981) employed a variant of the radial maze in which each arm was marked with distinctive visual and tactile cues, and the importance of extramaze cues that are presumed to support cognitive mapping was degraded. As in the usual version of the radial maze task, the animals had to avoid returning to already visited arms, but their choices had to be based on cues rather than places. Their results showed that rats that had learned the task preoperatively were unable to relearn after fimbria–fornix lesions, supporting the idea that hippocampal damage can profoundly impair working memory on a nonspatial task.

Olton and his colleagues have also presented evidence that damage to the hippocampal system does not seriously impair spatial mapping if the task does not have a working memory component as well. Olton and Papas (1979) trained rats on the version of the radial maze that permits simultaneous assessment of both working and reference memory. After receiving fimbria–fornix lesions, the animals displayed enduring deficits on the working memory component but only transient disruption of reference memory. In a subsequent experiment, Walker and Olton (1984) trained rats preoperatively to approach a central goal from three different directions. Following fimbria–fornix lesions there was a transient disruption of performance that quickly recovered to preoperative levels. The critical feature of the experiment was a transfer test on which the ability to approach the goal from the fourth direction was assessed. None of the rats had ever been tested for approach to the goal from this direction so accurate performance should depend upon the formation of a cognitive map of the experimental environment. Both control and lesioned subjects showed accurate performance on the transfer test, supporting the idea that spatial mapping does not require an intact hippocampal system.

Findings from other laboratories question this conclusion. Morris, Garrud, Rawlins, and O'Keefe (1982) found that rats with large hippocampal lesions were unable to acquire efficient place learning in the water maze as long as the platform remained hidden. If the platform was raised so that it was visible, the hippocampal group learned quickly, demonstrating that their ability for simple cue learning was not impaired. However, when the platform was again hidden the performance of the hippocampal group quickly deteriorated and did not recover.

On the same task, lesions of the CA3–CA4 fields (produced by microinjections of the neurotoxin kainic acid) or of the dentate gyrus (produced by microinjections of colchicine) produced a similar pattern of persistent deficits without recovery (Sutherland, Whishaw, & Kolb, 1983). These results are precisely as predicted by cognitive mapping theory because many place cells are located in the CA3 field that receives important inputs from the dentate gyrus. However, on other tasks such as the radial maze (Handelman & Olton, 1981), considerable behavioral recovery generally occurs with postoperative training after kainic acid-induced lesions. Differences in the task, extent of preoperative training or in the amount of damage to hippocampal cell fields could explain the differences in the duration of the impairment. The extent of damage to structures outside the hippocampus may also be important.

In a series of experiments Jarrard and his colleagues (Jarrard, 1983; Jarrard, Okaichi, Steward, & Goldschmidt, 1984) have provided a direct comparison of the working memory and cognitive mapping hypotheses. In the same radial maze they created both a place- and a cue-learning task by making extramaze or intramaze stimuli relevant, respectively. On both versions, some places or some cues were never baited so a reference memory as well as a working memory component existed for both the place and the cue task. After preoperative acquisition to criterion on both place and cue tasks, lesions that involved the entire hippocampus, various cell fields within the hippocampus, or connections to the hippocampus were produced. On postoperative tests, large hippocampal lesions (produced by aspiration) impaired both working and reference memory on the place task but had no significant effect on either working or reference memory on the cue task. Lesions of the fimbria–fornix or the entorhinal cortex impaired working memory on both place and cue tasks, and reference memory for the place but not the cue task.

The performance of rats with large hippocampal lesions is in accordance with cognitive mapping theory in that performance on the spatial but not the nonspatial version of the task was disrupted. Subsequent experiments suggested that this conclusion was premature. In these studies selective lesions were placed in the CA1–CA3 cell fields or in the dentate gyrus. Such lesions caused only a temporary effect on the place task (Jarrard, 1983; Jarrard, et al. 1984). Similar results were obtained in animals that sustained extensive damage to all hippocampal cell fields including the dentate gyrus produced by microinjections of the neurotoxin, ibotenic acid (Jarrard, 1985). These lesions would have destroyed most of the "place" cells, the putative components of the cognitive mapping system proposed by O'Keefe and Nadel (1978), but spared many of the fibers of passage destroyed by the large hippocampal lesions produced by aspiration.

Lesions of the subiculum (produced with ibotenic acid) also caused only temporary disruption of performance on the place task, but combined lesions of the hippocampal cell fields plus the subiculum caused persistent deficits in working memory for both place and cue tasks and in reference memory for the place task. In other words these lesions duplicated the effects of fimbria–fornix lesions.

The experiments of Jarrard's group are quite damaging to the working memory theory because none of their large or restricted lesions of the hippocampus or its connections produced selective impairments of working memory for either cue or place tasks. When such deficits were found, spatial reference memory was also impaired to an equivalent degree. Despite the fact that impairments were more common on place than on cue tasks, the cognitive mapping theory fares only a little better. Based on Jarrard's work complex spatial mapping can be accomplished by animals with massive damage to the "place" cells.

Jarrard's work directs attention away from the hippocampus proper to the subiculum and the entorhinal cortex as potentially more important components of the temporal lobe spatial memory circuitry.

Frontal Lobe

The idea that the frontal lobes play an important role in spatial memory has a long history. The original demonstration that primates with frontal lobe lesions are impaired on the spatial delayed response task but not immediate discrimination tasks (Jacobsen, 1935) has been replicated countless times in several different species (Fuster, 1980). Subsequent work showed that the largest deficits on delayed alternation as well as on the delayed response task resulted from damage to the sulcus principalis of the dorsolateral frontal cortex (Butters & Pandya, 1969; Butters, Pandya, Sanders, & Dye, 1971). Such lesions have little effect on nonspatial tasks such as object discrimination, reversal, or alternation (Mishkin, Vest, Waxler, & Rosvold, 1969). Both the temporal delay as well as the spatial nature of the task are critical to the deficit (Goldman, Rosvold, Vest, & Galkin, 1971).

It is possible to attribute the deficits on spatial delay tasks to defects in egocentric orientation. Pohl (1973) found that dorsolateral lesions interfered with spatial reversals, but not with landmark learning or reversal. On the landmark task a stimulus (the landmark) is placed between two marked spatial locations and the subject is rewarded for responding to the location that is closer to (or farther from) the landmark. Pohl (1973) suggested that accurate performance on spatial reversal as well as delayed response and alternation depends on memory for body position or direction of movement which in turn depends upon the integrity of the dorsolateral frontal cortex. Solution of the landmark task depends upon appreciating the spatial relationships among objects in extrapersonal space so no impairment would be expected since presumably the dorsolateral frontal cortex is not concerned with allocentric orientation.

Some studies with human patients are consistent with this proposal. The deficits on tasks that require imaginary rotation of one's own body (e.g., pointing to parts of their bodies that corresponded to those designated on a diagram) in human patients with frontal lobe damage coupled with their relatively normal performance on route following tasks or on tasks that require mental rotation of external objects (Butters et al., 1972; Semmes, Weinstein, Ghent, & Teuber, 1963) is consistent with the idea that frontal lesions impair egocentric but not allocentric orientation. Moreover, it is well established that testing procedures that permit subjects to maintain bodily orientation to the correct side during the delay alleviate the usual frontal deficits on spatial tasks (e.g., Kojima, Kojima, & Goldman-Rakic, 1982). Of course, these conditions greatly reduce the memory load so the "egocentric orientation" explanation somewhat begs the question.

Other findings are difficult to explain in terms of a defect in egocentric orientation. In the rat the medial frontal cortex is thought to be the analogue of the dorsolateral cortex in the primate (Kolb, 1984). After lesions of this area, rats are unable to learn to swim directly to the submerged platform in the Morris water maze (Kolb, Pittman, Sutherland, & Whishaw, 1982; Kolb, Sutherland, & Whishaw, 1983) although they find the platform quickly if it is not submerged (Whishaw & Kolb, 1984). These dramatic deficits were not observed if rats were trained prior to surgery. Lesions of the medial frontal cortex also impair performance on the radial maze, although the exact nature of the deficit is somewhat unclear. Becker et al. (1980) reported deficits in regaining accurate working memory (there was no reference memory component on the task) which eventually recovered, while Kolb et al. (1983) found impaired postoperative acquisition of the reference memory but not the working memory component.

Passingham (1985) attempted to create an analogue of the radial maze for monkeys. The animals were confronted with an array of 25 spatial locations and rewarded for the first "visit" to each location, but not for subsequent visits. Lesions of the sulcus principalis impaired performance, which Passingham attributed to a defect in working memory since the lesioned monkeys performed normally when there was no delay and none of the subjects appeared to employ a response strategy to solve the task. These results are difficult to reconcile with the idea that the dorsolateral cortex is exclusively concerned with egocentric space. Also inconsistent with the idea is the report of Milner, Ockleford, and Dewar (1977) that monkeys with dorsolateral frontal lesions were impaired in relearning on a preoperatively acquired landmark task, similar to the one used by Pohl (1973). The defective performance of their animals was almost entirely the result of an increase in perseverative errors, a frequent consequence of frontal lobe damage.

Patients with unilateral frontal lobe lesions perform poorly on tasks that appear to depend more on allocentric than on egocentric orientation. Such patients learn stylus mazes more slowly than normal subjects (Canaran, 1983; Corkin, 1965; Milner, 1965), but it is not clear that these deficits reflected disturbed spatial memory. Unlike patients with temporal or parietal lobe lesions (whose maze learning is also impaired), patients with frontal lesions frequently defied the rules of the problem by failing to respond appropriately to the error signal. Furthermore, Milner et al. (1977) found no impairment in relearning a preoperatively acquired stylus maze in monkeys with dorsolateral frontal lesions. Since Smith and Milner (1981) found that patients with frontal but not temporal lobe lesions could recall the location of objects normally even after a 24 hour delay, a general defect in spatial memory seems unlikely. Instead the maze learning deficits may reflect more general cognitive impairments in controlling sequential responding, perhaps due to the perseverative errors characteristic of frontal lobe patients (Brody & Pribram, 1978; Petrides & Milner, 1982) or in the temporal integration of behavioral sequences (Fuster, 1980). The latter suggestion is strongly supported by the finding that changing the temporal structure of the standard delayed alternation task leads to dramatic improvement in the performance of monkeys with frontal lesions. In the usual version of this task, the intertrial interval is fixed and frontal monkeys do very badly. But when 5- and 15-second-long intervals were alternated the monkeys performed normally (Pribram, Plotkin, Anderson & Leong, 1977; Pribram & Tubbs, 1967). Changing the temporal structure of the task apparently made it comprehensible to the brain-damaged monkeys, and eliminated their spatial memory deficit. In summary, the memory deficits seen in frontal lobe damage patients may reflect a "forgetting to remember" disorder, as patients may be able to learn new material in normal fashion, but show impairment if interference is imposed between learning and recall (Cermak & Butters, 1972). The deficits may thus reflect disturbances in attention, preservation and lack of initiative (Stuss & Benson, 1984).

Posterior Neocortex

Damage to the posterior neocortex is associated with a great variety of visuoperceptual and visuospatial defects, which include defects in localizing points in space, judging directional orientation, depth perception, and hemispatial neglect. Most of these disorders are more common in patients with lesions of the right hemisphere (Benton, 1985; Heilman, Watson, & Valenstein, 1985).

Parietal lobe lesions impair the ability of humans to follow routes (Ratcliff & Newcombe, 1973; Semmes et al., 1963) and perform reversible operations on objects in space (Butters & Barton, 1970). These observations were originally interpreted to support the view that the parietal lobe was especially concerned with allocentric or extrapersonal space (e.g., Butters et al., 1972) but relatively unimportant for egocentric orientation. The finding that monkeys with parietal lesions were impaired on a landmark task (described above) but not on spatial reversals (Pohl, 1973) was interpreted as supporting this view. However, Ratcliff (1979) observed deficits in patients with right posterior lesions on a test that required mental rotations of the patients' own bodies. Clearly, posterior neocortical damage can disturb egocentric localization as well.

The bewildering variety of disorders of spatial perception associated with posterior neocortical lesions makes interpretation of the effects of such lesions on learning and memory very difficult. For example, De Renzi, Faglioni, and Previdi (1977) studied spatial memory using the Corsi block test. On this task, the examiner touches a number of blocks from an array in a particular sequence that the subject must reproduce. Patients with unilateral lesions of the posterior neocortex on either side were impaired in learning a sequence that was two blocks longer than their immediate memory span. The largest effects were seen if the lesions produced visual field defects, but it is difficult to interpret the data because the immediate spatial memory span was also reduced. To circumvent this problem, De Renzi et al. (1977) used a short sequence that was within every patients' immediate memory span. Defects in memory were evident in patients with right hemisphere lesions at retention intervals as short as 8 seconds, but the effects were reliable only if the lesions produced visual field defects. A word of caution is in order here as Smirni, Villardita, and Zappala (1983) found that spatial memory performance on the Corsi block test depends not only on the path length but also the particular sequence. The findings of De Renzi et al. (1977) are thus difficult to interpret because the memory task might not have been within every patient's immediate memory span. Future research may need to address the question of which particular sequences (configurations) are most sensitive in detecting impairments.

Acquisition of accurate performance on visual and tactual stylus mazes is impaired by posterior neocortical lesions (Corkin, 1965; Milner, 1965). The deficits were greater if the damage involved the right hemisphere than the left (Ratcliff & Newcombe, 1973). Because deficits in maze learning were not correlated with performance on the Mooney faces test (Newcombe & Russell, 1969), visuoperceptual difficulties are probably not responsible for the learning defects. Additional support for this conclusion comes from case M.A. reported by De Renzi, Faglioni, and Villa (1977). This patient sustained an injury to the right temporoparietal region after a cerebrovascular accident. Although she performed normally on several tests of general intelligence, verbal learning and memory, and visual and tactual perception, she had great difficulty finding her way in both familiar and unfamiliar places and she was unable to learn the visual stylus maze despite extensive training.

Although the performance of patient M.A. suggests that deficits on tests of anterograde spatial learning may be correlated with retrograde defects in topographic memory, very few attempts to measure deficits in remote spatial memory can be found in the literature. Notable exceptions are the studies of geographical memory of Benton, Levin, and Van Allen (1974) and of Morrow, Ratcliff, and Johnston (1985). Benton et al. (1974) found that although patients with unilateral lesions of either hemisphere had little difficulty naming the state in which a city was located, they were impaired in stating the direction of travel from one place to another and in accurately locating places on an outline map of the U.S. Equally severe deficits were observed in patients with left- and right-hemisphere damage, but on the localization task both patient groups shifted the location of places toward the side of the lesion. Thus, hemispatial neglect may have contributed to the memory deficit. A problem of some concern is the confounding of spatial memory tasks (such as topographic memory for the U.S. map) with representation of enabling skills (e.g., overt constructional responses, visual scanning, or unilateral neglect). Morrow, Ratcliff, and Johnston (1985) have recently shown that when these enabling skills are not required, patients with right hemisphere (RH) lesions show an inability to manipulate interval representations of memory. More specifically, they fail to estimate distances between U.S. cities on a "cognitive map" as accurately as control subjects, which suggests impaired allocentric spatial memory. Unfortunately the site of the lesions was not specified, nor was it possible to determine at which stage of spatial processing disruption occurred or if egocentric memory was also impaired. Clearly, however, the deficits were not due to an inability to estimate distances per se, nor due to lack of knowledge of where the cities are located as the patients with RH lesions did as well as the control subjects on estimating distances between symbols as opposed to cities, and on locating cities by placing *x*'s on an outline map of the U.S.

Caudate Nucleus

Lesions within the caudate nucleus produce a pattern of defects on spatial tasks that generally resemble those that accompany damage to the prefrontal cortex. Thus, damage to the caudate impairs performance on delayed response and delayed alternation tasks (Battig, Rosvold, & Mishkin, 1960). Lesions in the anterodorsal portion of the head of the caudate, which receives projections from the dorsolateral frontal cortex, produce the most serious deficits on spatial tasks, but they do not affect visual discrimination learning or object reversals (Divac, Rosvold, & Szwarcbart, 1967).

Patients with Huntington's disease (HD), a degenerative disease that initially strikes the caudate nucleus, are impaired on delayed alternation (Oscar-Berman, Zola-Morgan, Oberg, & Bonner, 1982). In the same study, the HD patients were slightly impaired on a delayed response task, but the effect was not significant, perhaps because of a ceiling effect.

Potegal (1972) has proposed that the caudate nucleus is involved in processing sensory information (especially vestibular inputs) that are important for egocentric orientation. In an early experiment with rats, Potegal (1969) found that lesions in the caudate nucleus disrupted learning a spatial discrimination that could only be solved by turning 30 degrees away from straight ahead, but had no effect on the learning of an equally difficult task for which the correct solution was based on relative position. Successful solution of the first task required the rat to use its body as a frame of reference. No such egocentric requirement existed on the latter task, which presented the rat with a difficult allocentric discrimination. Subsequently Abraham, Potegal, and Miller (1983) examined the performance of rats with lesions of the caudate nucleus or hippocampus on a return from passive transport task. On this task the animal was required to return to the correct place in the environment after being passively transported away from the goal in an opaque cart through a 90 degree turn. As successful performance on the task depends on proprioceptive and kinesthetic cues, the task qualifies as a measure of egocentric orientation. As predicted, lesions of the caudate nucleus, but not the hippocampus, disrupted performance, supporting the idea that the caudate nucleus and hippocampus are involved in different aspects of spatial behavior. However, the hippocampal lesions were quite small and no test of allocentric spatial behavior was included. Hence, no double dissociation was established. However, large hippocampal, but not caudate, nucleus lesions, do impair performance on the radial maze (Becker et al., 1980).

The involvement of the caudate nucleus in egocentric orientation is supported by studies of spatial behavior in patients with HD. Potegal (1971) required subjects to view a target and then point to it while blindfolded. Huntington's disease patients performed normally except when they had to respond after stepping sideways to the left or right. Under this condition, which is a human analogue of the return from passive transport task described above, the HD patients were impaired. In a subsequent experiment Brouwers, Cox, Martin, Chase, and Fedio (1984) found that HD patients were impaired on a road map test that required them to perform a mental rotation of their own bodies, a test of egocentric orientation. The HD patients were not significantly impaired in learning a visual stylus maze or in copying the Rey-Osterreith Complex figure. A comparison group of patients with Alzheimer's disease exhibited the opposite pattern of deficits.

Although patients with HD succeed on some allocentric tasks, they fail others. Visuospatial deficits on tests of visual memory and visuospatial judgment appear early in the course of HD (Josiassen, Curry, & Mancall, 1983). We recently developed a test of geographical memory that requires subjects to locate cities and other geographical features on outline maps of regions that typically consist of 2–4 continguous U.S. states. The test was developed as a means of measuring remote memory for spatial information (Beatty, 1985). Huntington's disease patients and control subjects were asked to locate places on maps of the region of current residence as well as of the region in which they were born and raised. The HD patients were equally impaired on both tasks.

We also developed an anterograde test of map learning which was similar in format to the geographical memory test. On this task subjects had to learn the location of 15 hypothetical "towns" located in three imaginary "states." The HD patients were profoundly impaired on this test as well (Beatty, unpublished observations.)

Medial Diencephalon

The importance of the medial diencephalon for memory has long been presumed because of the consistent association of pathology in this region of the brain with the amnesia exhibited by patients with alcoholic Korsakoff's syndrome (AK). Historically debate has focused on whether the mammillary bodies (MMB) or the mediodorsal nucleus of the thalamus (MD) is responsible for the amnesia. Victor, Adams, and Collins (1971) concluded that damage to MD was necessary and sufficient to the memory disturbance in AK, but as others (Butters, 1984; Markowitsch, 1982) have pointed out, the data presented by Victor et al. (1971) are also amenable to the interpretation that damage to both the MD and the MMB is required for severe and lasting memory disturbance.

More recent studies with computed temography scans indicate that chronic abuse of alcohol (regardless of whether or not AK develops) is associated with considerable cortical atrophy, particularly of the frontal lobes (Wilkinson, 1982). Neuropsychological studies (Squire, 1982) suggest that at least some of the cognitive changes in AK may be more closely associated with frontal lobe than with diencephalic damage.

Patients with AK are impaired on delayed response, delayed alternation, and spatial reversals (Oscar-Berman & Zola-Morgan 1980a; Oscar-Berman et al., 1982), but it seems unlikely that these deficits can be attributed to damage to any single diencephalic structure. Lesions of both the MD (Isseroff, Rosvold, Galkin, & Goldman-Rakic, 1982) and the MMB (Holmes, Jacobson, Stein, & Butters, 1983) can impair delayed alternation in monkeys, but the effects are relatively modest compared to the effects of lesions of the dorsolateral frontal cortex. For example, MMB lesions impaired performance only if a 30-second delay was imposed between responses. Performance was normal with a 5-second delay, a condition that is more than adequate to produce a profound deficit in monkeys with frontal lobe lesions or in AK patients (Goldman et al., 1971; Oscar-Berman et al., 1982). Furthermore, Isseroff et al. (1982) did not observe a reliable impairment on the delayed response task after MD lesions, although there was a trend toward a deficit. More extensive lesions of the MD do produce deficits on delayed response (Schulman, 1964), but these lesions damaged other thalamic areas as well. Deficits on spatial reversals have generally been observed after MD lesions in rats (Kolb, 1977; Means, Hershey, Waterhouse, & Lane, 1975) and monkeys (Holmes, Butters, Jacobson, & Stein, 1983), but deficits on this task are observed after damage to a great many brain regions so this finding is not particularly informative.

The limited data available suggest that diencephalic structures are not particularly important for spatial mapping. Kolb, Pittman, Sutherland, and Whishaw (1982) found that MD lesions in rats caused no impairment in learning the radial maze or the Morris water maze. Kesler, Markowitsch, and Otto (1982) observed a small deficit on the radial maze in rats with MD lesions, but only if a 1-hour delay was imposed during the run.

Jarrard et al. (1984) found no effect of MMB lesions on the radial maze. These negative results contrast strikingly with the large and persistent deficits produced by lesions of the frontal cortex and hippocampal formation on these tasks. In summary, it appears that the negative findings in animal studies would question the wisdom of ascribing the cognitive deficits seen in AK entirely to insults to the medial diencephalon. Neuropathological and neuropsychological data suggest considerable frontal lobe pathology in these patients. In addition, the hippocampus may undergo pathological changes as well (Victor et al., 1971). Animal studies indicate that damage to the frontal cortex and hippocampus may play an important role in spatial memory deficits of AK patients.

Neuropharmacology

A strong case can be made for the importance of central cholinergic circuits in spatial memory. Treatment with centrally active cholinergic antagonists such as atropine or scopolamine leads to consistent impairment, but peripherally active forms of these drugs that penetrate the brain very poorly are ineffective.

Acquisition of accurate working memory and reference memory on the radial maze (Stevens, 1981; Watt, Stevens, & Robinson, 1981) and the Morris water maze (Sutherland, Whishaw, & Regehr, 1982) are impaired by treatment with anticholinergic drugs. In subjects that have already acquired stable performance on a task, deficits are uncommon on the reference memory components except at high doses. Thus, on the radial maze scopolamine impairs working memory but spares reference memory (Beatty & Bierley, 1985; Wirsching, Beninger, Jhamandas, Boegman, & El-DeFrawy, 1984). On the Morris water maze atropine impairs acquisition, but has no effect on memory for a previously learned spatial location (Whishaw, 1985). The acquisition deficit caused by anticholinergic drugs was specific to the allocentric spatial task. Acquisition of cue and position habits was not affected (Whishaw, 1985). In monkeys, scopolamine impairs performance on the spatial delayed response problem (Bartus & Johnson, 1976). The deficit is negligible if the animals are allowed to respond immediately and increases as the length of the delay increases.

The influence of anticholinergic drugs on spatial memory appears to arise from an action on some aspect of encoding rather than upon a memory storage or consolidation like process. To establish this point we made use of the fact that in rats tested in the radial maze spatial working memory remains highly accurate if a delay of several hours is imposed during the test (Beatty & Shavalia, 1980). With a 5-hour delay imposed during the run, doses as high as 5 mg/kg of body weight had no effect on retention when given immediately after the rats completed their first four or eight choices (Godding, Rush, & Beatty, 1982). This dose is twenty times the dose required to disrupt spatial working memory if given just prior to testing.

The effects of anticholinergic drugs on spatial memory most likely result from blockade of cholinergic inputs to the hippocampus and frontal cortex. Recent evidence indicates that selective lesions of either the medial septal nucleus (the origin of cholinergic inputs to the hippocampus), or of the nucleus basalis (the origin of cholinergic inputs to the cerebral cortex) interfere with accurate spatial memory (Hepler, Olton, Wenk, & Coyle, 1985; Murray & Fibiger, 1985; Salamone, Beart, Alpert, & Iversen, 1984). The pattern of deficits produced by these lesions closely resembles the influence of anticholinergic drugs on the behavior of intact animals. If lesions are made before training then acquisition of spatial behavior is impaired. If lesions are made after training, either working (Knowlton, Wenk, Olton, & Coyle, 1985) or reference memory (Kesner, DiMattia, & Crutcher, 1987) may be disrupted.

Both the cerebral cortex and the hippocampus also receive important noradrenergic inputs that arise from the locus caeruleus. However, destruction of these inputs with neurotoxins has no effect on acquisition of spatial behavior in the radial maze or the Morris water maze (Chrobak, De Haven, & Walsh, 1985; Harrell, Barlow, Miller, Haring, & Davis, 1984; Hagen, Alper, Morris, & Iversen, 1983; Sutherland, Kolb, Whishaw, & Becker, 1982). Likewise administration of noradrenergic antagonists does not affect working memory in well trained rats (Beatty & Rush, 1983).

Aging

With very few exceptions, study of spatial memory in humans reveal that older people learn and remember spatial information more poorly than young adults. Age-related declines in accuracy have been observed on a number of tasks, including memory for the location of buildings on a map (Perlmutter, Metzger, Nezworski, & Miller, 1981), objects arranged in a set of cubicles (Waddell & Rogoff, 1981), objects or their verbal labels arranged on a 6 × 6 matrix (Pezdek, 1983), and chess pieces on a chess board (Charness, 1981). Comparable deficits have been observed when subjects were told to remember the spatial locations of structures on a map (intentional learning) as well as when they were not so instructed (incidental learning) (Light & Zelinski, 1983).

Deficits in acquiring and remembering spatial information have consistently been observed in aged rats. Impairments have been reported on tasks that principally require reference memory such as the Morris water maze (Barnes, 1979; Gage, Dunnett, & Bjorklund, 1984) as well as on the version of the radial maze that measures only working memory (Barnes, Nadel, & Honig, 1980; Davis, Idowu, & Gibson, 1983; de Toledo-Morrell, Morrell, & Fleming, 1984; Wallace, Krauter, & Campbell, 1980.) Old rats are also impaired in acquiring the multiple-unit T maze (Goodrick, 1968), a task that does not require spatial mapping. It is important to note that there are large individual differences within the aged groups. Though the majority of old animals are impaired in spatial memory, a substantial minority are not.

Most attention has been directed at the possibility that changes in the hippocampus might underlie defective spatial memory in aged animals. The likely importance of this region is supported by the recent observations that although glucose utilization was reduced in 14 of 45 brain regions examined, in only 5 of the 14 regions was the level of metabolic activity correlated with accuracy of performance on the Morris water maze (Gage, Kelly & Bjorklund, 1984). The five areas included four regions within the septohippocampal system and the prefrontal cortex.

Several anatomic and physiologic changes have been described in the hippocampus of aging rats that might contribute to the deficits in spatial memory. These include a loss of synapses, particularly in the dentate gyrus (Bondareff & Geinisman, 1976), and a reduction in neuronal cell density (Landfield, Rose, Sandles, Wohlstader, & Lynch, 1977). These anatomic changes may relate to the loss of specificity of the hippocampal "place" cells of aged animals described by Barnes, McNaughton, and O'Keefe (1983).

Loss of muscarinic receptors (Lippa, Pelham, Beer, Critchett, Dean, & Bartus, 1980) and reduced responsiveness to iontophoretically injected acetylcholine (Segal, 1982) also occur in the hippocampus of aged rats. These changes may be important because much evidence (described earlier) indicates that pharmacological blockade of muscarinic receptors interferes with spatial working memory. Furthermore, treatment with the drug 3, 4-diaminopyridine (which increases acetylcholine release) improves spatial memory in old rats (Davis et al., 1983; Peterson & Gibson, 1983).

Failing spatial memory in aged rats has been correlated with changes on two neurophysiological markers, kindling and long-term enhancement, which are thought to provide neuronal models of long-term memory. In the kindling paradigm, repeated exposure to subthreshold electrical stimulation that initially elicits only a local afterdischarge ultimately elicits a major motor seizure. Once established, the kindled seizure can be elicited after many months without stimulation. Long-term enhancement refers to an increase in synaptic activity that can be produced by repetitive stimulation of many of the excitatory pathways in the hippocampus. Once enhancement is produced, it persists for days or weeks.

Overall, kindling is more easily induced and enhancement is longer maintained in young animals. More important is the fact that both ease of kindling and persistence of long-term enhancement are closely correlated with accuracy of spatial memory among aged rats (Barnes, 1979; de Toledo-Morrell, Morrell, & Fleming, 1984). Furthermore, pentoxifylline, a drug that improves spatial memory in the radial maze in old rats (de Toledo-Morrell, Morrell, Fleming, & Cohen, 1984), also reduces the number of stimulations required to produce kindling and prolongs the duration of long-term enhancement.

Though these observations offer the exciting prospect that age-related deficits in spatial memory (and perhaps memory in general) may be associated with pathology in the hippocampus, a note of caution is in order. Jarrard and his colleagues (reviewed earlier in the section on Temporal Lobe) have recently shown that lesions that destroy most of the cells in the dentate gyrus or the CA1–CA3 cell fields can be destroyed with little effect on spatial memory. These lesions destroyed most of the "place cells" and most of the synaptic sites that undergo long-term enhancement in the studies of Barnes (1979) and de Toledo-Morrell, Morrell, & Fleming (1984).

It is important to note that Jarrard's animals were trained prior to surgery. Although no data are available on subjects with comparable lesions trained postoperatively, anterograde memory is usually more severely affected than retrograde memory by most types of brain damage. For example, lesions of the hippocampal CA1 field interfere with acquisition of accurate working memory in the radial maze but have not effect on postoperative retention (Jarrard, 1978).

For obvious reasons, most studies of aging and spatial memory use cross-sectional comparisons of aged and younger groups on measures of new learning. As learning occurs after age-related declines in brain function have begun, the deficits displayed by the aged subjects are analogous to anterograde memory defects after more acute brain injury.

We have recently found that deficits in spatial working memory in the radial maze are not observed in aged rats that received extensive experience on the task starting when they were young (Beatty, Bierley, & Boyd, 1985). Such animals exhibit almost no loss of accurate working memory even after 10 months without practice (Bierley, Rixen, Tröster, & Beatty, 1986). Control experiments indicated that the old rats were not employing a response strategy or converting the working memory task to a reference memory problem. Thus, the capacity to display accurate spatial working memory was, in fact, preserved. Because no physiological or anatomic measures were taken, it is not yet known whether the prior training which preserved accurate spatial memory altered the course of aging of the hippocampus. If it did not, then aged-related changes in hippocampal function may be important only for remembering spatial information acquired when the individual is already aged.

Summary and Conclusions

We have examined the effects of lesioning various cortical areas and subcortical structures, as well as the effects of anticholinergics and aging, on spatial memory. Although the classification of spatial behavior as egocentric or allocentric is commonly made, the distinction is difficult to support at times in the case of nonhuman subjects. Even in the case of humans it may be premature to speculate that distinct neural pathways underlie these spatial behaviors as they may share, to some extent, a common neural substrate. A clear statement regarding allocentricity versus egocentricity is further complicated by the fact that virtually no studies control for the possibility that a subject uses some strategy to convert an egocentric spatial task to an allocentric one or vice versa. Unless a necessary pathway common to both allocentric and egocentric spatial problem solving is disrupted (assuming such a pathway exists), it would theoretically be possible to find no deficit in spatial problem solving if the subject compensated by using a strategy to convert the problem to some other form requiring a different neural substrate.

Alternatively, if the pathways subserving egocentric and allocentric spatial behaviors are partially shared, deficits on both egocentric and allocentric tasks might arise. In general, the experimental evidence supports this latter possibility. Thus damage to the caudate nucleus as in patients with HD or to the prefrontal cortex in rats leads to deficits in allocentric as well as egocentric spatial tasks. Likewise, posterior neocortical lesions can disturb egocentric as well as allocentric spatial functions. These findings may question the usefulness of the egocentric–allocentric dichotomy as a guide for future studies aimed at uncovering the neural circuitry controlling spatial behavior. If the brain respects the egocentric–allocentric distinction, studies that the more refined behaviorally and anatomically will be required to demonstrate this point.

In the case of temporal lobe lesions, it is fairly well established that right hemisphere lesions are far more likely to result in spatial learning and memory deficits than left hemisphere lesions and that the magnitude of the deficit is proportional to the extent of damage, especially the extent of hippocampal damage. Although the evidence from rats clearly supports the view that the hippocampal damage is critical to the anterograde spatial memory impairment, studies with nonhuman primates suggest such a conclusion is premature. Either amygdaloid or hippocampal lesions can impair delayed alternation performance, but impairments on the delayed response task may require lesions involving both structures. Such combined amygdalohippocampal lesions produce a general loss of recognition memory (Mishkin, 1978; Zola-Morgan & Squire, 1985a). Thus there is no reason to suppose that the temporal lobe is specialized to process spatial as opposed to nonspatial memories. The same conclusion applies to the mediodorsal thalamus. Although lesions in this region can produce deficits in tests of spatial memory, they also disrupt recognition memory in primates (Aggleton & Mishkin, 1983; Zola-Morgan & Squire, 1985b) and cause learning deficits on a broad range of tasks in rats (Delacour, 1971; Winocur, 1985).

In a recent paper, Pandya and Yeterian (1984) suggest that the parieto–temporal region of the posterior neocortex plays a critical role in spatial memory. Specifically, they propose that two different output pathways from the inferior parietal lobule (area 7) to the hippocampus subserve different aspects of spatial memory formation. The more rapidly acting pathway receives inputs from area 7 and the cingulate gyrus via the presubiculum and conveys information about the spatial location of simple objects whose localization has immediate survival value for the organism. The more slowly acting pathway conveys information from area 7 and the cingulate gyrus to the hippocampus and the amygdala via the parahippocampal gyrus. This pathway is presumed to be involved in the formation of more complex spatial images (e.g., mazes).

Newcombe (1985) has pointed out that the Pandya and Yeterian (1984) model can be interpreted as providing the anatomic basis for separate processing of short- and long-term spatial memories. She describes a patient with damage to the supramarginal gyrus and the superior parietal lobule including the underlying white matter of the right hemisphere. The patient exhibited markedly impaired maze learning, but normal performance on the Corsi block span test and normal copying and memory of complex figures. This patient's selective deficit in long-term spatial memory contrasts with two cases described by De Renzi and Nichelli (1975). These patients displayed normal maze learning but reduced spatial memory span on the block tapping test. Though there is ample evidence that deficits on tests of spatial memory can be dissociated from gross defects in attention and perception in some patients with parietal lobe injury, it is unclear whether their impairments are better described as defects in spatial information processing or spatial memory.

Damage to the dorsolateral prefrontal cortex causes severe impairments in tasks that involve spatial memory, such as the delayed response and delayed alternation problems. Although the spatial nature of the test is critical to the deficit, changing the temporal structure of the task greatly improves performance by subjects with damage to the prefrontal cortex. Hence the impairment does not seem to involve a failure a spatial memory so much as a more general defect in the sequential control of behavior.

The caudate nucleus may be involved in anterograde and retrograde spatial memory (especially egocentric), but its role may not be direct but rather attributable to its role in processing sensory and vestibular information.

In summary, it can be stated that although several cortical and subcortical regions are involved in spatial memory, their specific roles have not been elucidated. Future studies may fruitfully resort to sequential lesioning experiments. However, considerable attention will need to be given to comparability across studies (e.g., using similar species, similar tasks, and similar lesions). Effort also needs be directed at establishing the role of neural structures in retrograde spatial memory, an area of investigation grossly neglected up to this point. Whatever area of investigation is addressed, relatively "pure" spatial tasks will need to be used (Morrow et al., 1985). Once such findings are forthcoming, it will be easier to establish which, if any, neural structures exclusively subserve spatial memory, or whether spatial memory is served by structures and pathways serving other memory functions as well.

References

Abraham, L., Potegal, M., & Miller, S. (1983). Evidence for caudate nucleus involvement in an egocentric task: Return from passive transport. *Physiological Psychology*, *11*, 11–17.

Aggleton, J.P., & Mishkin, M. (1983). Visual recognition impairment following medial thalamic lesions in monkeys. *Neuropsychologia*, *21*, 189–197.

Barnes, C.A. (1979). Memory deficits associated with senescence: A behavioral and electrophysiological study in the rat. *Journal of Comparative and Physiological Psychology*, *93*, 74–104.

Barnes, C.A., McNaughton, B.L., & O'Keefe, J. (1983). Loss of place specificity in hippocampal complex spike cells of sensecent rats. *Neurobiology of Aging*, *4*, 113–119.

Barnes, C.A., Nadel, L., & Honig, W.K. (1980). Spatial memory deficit in sensecent rats. *Canadian Journal of Psychology*, *34*, 29–39.

Bartus, R.T., & Johnson, H.R. (1976). Short-term memory in the rhesus monkey: Disruption from the anti-cholinergic scopolamine. *Pharmacology, Biochemistry and Behavior*, *5*, 39–46.

Battig, K., Rosvold, H.E., & Mishkin, M. (1960). Comparison of frontal and caudate lesions on delayed response and delayed alternation in monkeys. *Journal of Comparative and Physiological Psychology*, *53*, 400–404.

Beatty, W.W. (1985). Assessing remote memory for space: The Fargo Map Test. *Journal of Clinical and Experimental Neuropsychology*, *7*, 640.

Beatty, W.W., & Bierley, R.A. (1985). Scopolamine degrades working memory but spares reference memory in the radial maze: Dissimilarity of anticholinergic effect and restriction of distal visual cues. *Pharmacology, Biochemistry and Behavior*, *23*, 1–6.

Beatty, W.W., Bierley, R.A., & Boyd, J.G. (1985). Preservation of accurate spatial memory in aged rats. *Neurobiology of Aging*, *6*, 219–225.

Beatty, W.W., MacInnes, W.D., Porphyris, H. Tröster, A.I., & Cermak, L.S. (in press). Preserved topographical memory following right temporal lobectomy. *Brain and Cognition*.

Beatty, W.W., & Rush, J.R. (1983). Spatial working memory in rats: Effects of

monoamine antagonists. *Pharmacology, Biochemistry and Behavior*, *18*, 7–12.
Beatty, W.W., & Shavalia, D.A. (1980). Spatial memory in rats: Time course of working memory and effect of anesthetics. *Behavioral and Neural Biology*, *28*, 454–462.
Becker, J.T., Walker, J.A., & Olton, D.S. (1980). Neuroanatomical basis of spatial memory. *Brain Research*, *200*, 307–320.
Bengelloun, W.A., Nelson, D.J., Gerth, J.M., & Beatty, W.W. (1975). Variations in magnitude of reward and position reversal learning following septal lesions in the rat. *Physiological Psychology*, *3*, 240–244.
Benton, A. (1985). Visuoperceptual, visuospatial, and visuoanstructive disorders. In K.M. Heilman & E. Valenstein (Eds.). *Clinical Neuropsychology* (2nd ed.) New York: Oxford.
Benton, A.L., Levin, H.S., & Van Allen, M.W. (1974). Geographic orientation in patients with unilateral cerebral disease. *Neuropsychologia*, *12*, 183–191.
Bierley, R.A., Kesner, R.P., & Novak, J.M. (1983). Episodic long term memory in the rat: Effects of hippocampal stimulation. *Behavioral Neuroscience*, *97*, 42–48.
Bierley, R.A., Rixen, G.J., Tröster, A.I., & Beatty, W.W. (1986). Preserved spatial memory in old rats survives ten months without training. *Behavioral and Neural Biology*, *45*, 223–229.
Bondareff, W., & Geinisman, Y. (1976). Loss of synapses in the dentate gyrus of the senescent rat. *American Journal of Anatomy*, *145*, 129–136.
Brody, B.A., & Pribram, K.H. (1978). The role of frontal and parietal cortex in cognitive processing: Tests of spatial and sequential functions. *Brain*, *101*, 607–633.
Brouwers, P., Cox, C., Martin, A., Chase, T., & Fedio, P. (1984). Differential perceptual–spatial impairment in Huntington's and Alzheimer's dementias. *Archives of Neurology*, *41*, 1073–1076.
Butters, N. (1984). Alcoholic Korsakoff's Syndrome: An update. *Seminars in Neurology*, *4*, 226–244.
Butters, N., & Barton, M. (1970). Effect of parietal lobe damage on performance of reversible operations in space. *Neuropsychologia*, *8*, 205–214.
Butters, N., & Pandya, D. (1969). Retention of delayed-alternation: Effect of selective lesions of sulcus principalis. *Science*, *165*, 1271–1273.
Butters, N., Pandya, D., Sanders, K., & Dye, P. (1971). Behavioral deficits in monkeys after selective lesions within the middle third of the sulcus principalis. *Journal of Comparative and Physiological Psychology*, *76*, 8–14.
Butters, N., Soeldner, C., & Fedio, P. (1972). Comparison of parietal and frontal lobe spatial deficits in man: Extrapersonal vs. personal (egocentric) space. *Perceptual and Motor Skills*, *34*, 27–34.
Canaran, A.G.M. (1983). Stylus-maze performance in patients with frontal-lobe lesions and relationship to verbal and spatial abilities. *Neuropsychologia*, *21*, 375–382.
Cermak, L.S., & Butters, N.M. (1972). The role of interference and encoding in the short-term memory of Korsakoff patients. *Neuropsychologia*, *10*, 89–95.
Charness, N. (1981). Visual short-term memory and aging in chess players. *Journal of Gerontology*, *36*, 615–619.
Chrobak, J.J., De Haves, D.L., & Walsh, T.J. (1985). Depletion of brain norepinephrine with DSP-4 does not alter acquisition or performance of a radial-arm maze task. *Behavioral and Neural Biology*, *44*, 144–150.
Cooper, A., & Marshall, P. (1985). Spatial location judgments as a function of

intention to learn, and mood state: An evaluation of an alleged automatic encoding operation. *American Journal of Psychology*, *98*, 261–269.

Corkin, S. (1965). Tactually guided maze learning in man: Effects of unilateral cortical excisions and bilateral hippocampal lesions. *Neuropsychologia*, *3*, 339–351.

Corkin, S. (1984). Lasting consequences of bilateral medial temporal lobectomy: Clinical course and experimental findings in H.M. *Seminars in Neurology*, *4*, 249–259.

Davis, H.P., Idowu, A., & Gibson, G.E. (1983). Improvement of 8-arm maze performance in aged Fischer 344 rats with 3,4-diaminopyridine. *Experimental Aging Research*, *9*, 211–214.

De Renzi, E., Faglioni, P., & Previdi, P. (1977). Spatial memory and hemispheric locus of lesion. *Cortex*, *13*, 424–433.

De Renzi, E., Faglioni, P., & Villa, P. (1977). Topographical amnesia. *Journal of Neurology, Neurosurgery, and Psychiatry*, *40*, 498–505.

De Renzi, E., & Nichelli, P. (1975). Verbal and non-verbal short-term memory impairment following hemispheric damage. *Cortex*, *11*, 341–354.

de Toledo-Morrell, L., Morrell, F., & Fleming, S. (1984). Age-dependent deficits in spatial memory are related to impaired hippocampal kindling. *Behavioral Neuroscience*, *98*, 902–907.

de Toledo-Morrell, L., Morrell, F., Fleming, S., & Cohen, M.M. (1984). Pentoxifylline reverses age-related deficits in spatial memory. *Behavioral and Neural Biology*, *42*, 1–9.

Delacour, J. (1971). Effects of medial thalamic lesions in the rat: A review and an interpretation. *Neuropsychologia*, *9*, 157–174.

Divac, I., Rosvold, H.E., & Szwarcbart, M.K. (1967). Behavioral effects of selective ablation of the caudate nucleus. *Journal of Comparative and Physiological Psychology*, *63*, 184–190.

Fuster, J. (1980). *The Prefrontal Cortex: Anatomy, Physiology and Neuropsychology of the Frontal Lobe*, New York: Raven Press.

Gaffan, D. (1972). Loss of recognition memory in rats with lesions of the fornix. *Neuropsychologia*, *10*, 327–341.

Gage, F.H., Dunnett, S.B., & Bjorklund, A. (1984). Spatial learning and motor deficits in aged rats. *Neurobiology of Aging*, *5*, 43–48.

Gage, F.H., Kelly, P.A.T., & Bjorklund, A. (1984). Regional changes in brain glucose metabolism reflect cognitive impairments in aged rats. *Journal of Neuroscience*, *4*, 2856–2865.

Godding, P.R., Rush, J.R., & Beatty, W.W. (1982). Scopolamine does not disrupt spatial working memory in rats. *Pharmacology, Biochemistry and Behavior*, *16*, 919–923.

Goldman, P.S., Rosvold, H.E., Vest, B., & Galkin, T.W. (1971). Analysis of the delayed-alternation deficit produced by dorsolateral prefrontal lesions in the Rhesus monkey. *Journal of Comparative and Physiological Psychology*, *77*, 212–220.

Goodrick, C.L. (1968). Learning, retention, and extinction of a complex maze habit for mature-young and senescent Wistor albino rats. *Journal of Gerontology*, *23*, 298–304.

Gray, J.A., & McNaughton, N. (1983). Comparison between the behavioral effects of septal and hippocampal lesions. *Neuroscience and Biobehavioral Reviews*, *7*, 119–188.

Hagen, J.J., Alper, J.E., Morris, R.G.M., & Iversen, S.D. (1983). The effects of central catecholamine depletions on spatial learning in rats. *Behavioural Brain Research*, *9*, 83–104.

Handelman, G.E., & Olton, D.S. (1981). Spatial memory following damage to hippocampal CA3 pyramidal cells with kainic acid: Impairment and recovery following preoperative training. *Brain Research*, *217*, 41–57.

Harrell, L.E., Barlow, T.S., Miller, M., Haring, J.H., & Davis, J.N. (1984). Facilitated reversal learning of a spatial-memory task by medial septal injections of 6-hydroxydopamine. *Experimental Neurology*, *85*, 69–77.

Heilman, K.M., Watson, R.T., & Valenstein, E. (1985). Neglect and related disorders. In K.M. Heilman and E. Valenstein (Eds.), *Clinical Neuropsychology*, (2nd ed.), New York: Oxford.

Hepler, D.J., Olton, D.S., Wenk, G.L., & Coyle, J.T. (1985). Lesions in nucleus basalis magnocellularis and medial septal area of rats produce qualitatively similar memory impairments. *Journal of Neuroscience*, *5*, 866–873.

Holmes, E.J., Butters, N., Jacobson, S., & Stein, B.M. (1983). An examination of mammillary-body lesions on reversal learning sets in monkeys. *Physiological Psychology*, *11*, 159–163.

Holmes, E.J., Jacobson, S., Stein, B.M., & Butters, N. (1983). Ablations of the mammillary nuclei in monkeys: Effects on postoperative memory. *Experimental Neurology*, *81*, 97–113.

Isaacson, R.L., Nonneman, A.J., & Schmaltz, L.W. (1968). Behavioral and anatomical sequellae of damage to the infant nervous system. In R.L. Isaacson (Ed.), *The Neuropsychology of Development*. New York: Wiley.

Isseroff, A., Rosvold, H.E., Galkin, T.W., & Goldman-Rakic (1982). Spatial memory impairments following damage to the mediodorsal nucleus of the thalamus in rhesus monkeys. *Brain Research*, *232*, 97–113.

Jacobsen, C.F. (1935). Functions of the frontal association area in primates. *Archives of Neurology and Psychiatry*, *33*, 558–569.

Jarrard, L.E. (1975). Role of interference in retention by rats with hippocampal lesions. *Journal of Comparative and Physiological Psychology*, *89*, 400–408.

Jarrard, L.E. (1978). Selective hippocampal lesions: differential effect on performance by rats of a spatial task with preoperative versus postoperative training. *Journal of Comparative and Physiological Psychology*, *92*, 1119–1127.

Jarrard, L.E. (1983). Selective hippocampal lesions and behavior: Effects of kainic acid lesions on performance of place and cue tasks. *Behavioral Neuroscience*, *97*, 873–889.

Jarrard, L.E. (1985). Selective hippocampal lesions and behavior: Implications for current research and theorizing. In R.L. Isaacson & K. Pribram (Eds.), *The Hippocampus*. New York: Plenum.

Jarrard, L.E., Okaichi, H., Steward, O., & Goldschmidt, R.B. (1984). On the role of the hippocampal connections in the performance of place and cue tasks: Comparisons with damage to the hippocampus. *Behavioral Neuroscience*, *98*, 946–954.

Jones, B., & Mishkin, M. (1972). Lesions and the problem of stimulus–reinforcement associations. *Experimental Neurology*, *36*, 363–377.

Jones-Gotman, M. (1986a). Memory for designs: The hippocampal contribution. *Neuropsychologia*, *24*, 193–203.

Jones-Gotman, M. (1986b). Right hippocampal excision impairs learning and recall of a list of abstract designs. *Neuropsychologia*, *24*, 659–670.

Josiassen, R.C., Curry, L.M., & Mancall, E.L. (1983). Development of neuropsychological deficits in Huntington's disease. *Archives of Neurology*, *40*, 791–796.

Kesler, J., Markowitsch, H.J., & Otto, B. (1982). Subtle but distinct impairments of rats with chemical lesions in the thalamic mediodorsal nucleus tested in a radial arm maze. *Journal of Comparative and Physiological Psychology*, *96*, 712–720.

Kesner, R.P., DiMattia, B.V., & Crutcher, K.A. (1987). Evidence for neocortical involvement in reference memory. *Behavioral and Neural Biology*, *47*, 40–53.

Knowlton, B., McGowan, M., Olton, D.S., & Gamzu, E. (1985). Hippocampal stimulation disrupts spatial working memory even 8 hr after acquisition. *Behavioral and Neural Biology*, *44*, 325–337.

Knowlton, B.J., Wenk, G.L., Olton, D.S., & Coyle J.T. (1985). Basal forebrain lesions produce a dissociation of trial-dependent and trial-independent memory performance. *Brain Research*, *345*, 315–321.

Kojima, S., Kojima, M., & Goldman-Rakic, P.S. (1982). Operant analysis of memory loss in monkeys with prefrontal lesions. *Brain Research*, *248*, 51–59.

Kolb, B. (1977). Studies on the caudate-putamen and the dorsomedial thalamic nucleus of the rat: Implications for mammalian frontal-lobe functions. *Physiology and Behavior*, *18*, 237–244.

Kolb, B. (1984). Functions of the frontal cortex of the rat: A comparative review. *Brain Research Reviews*, *8*, 65–98.

Kolb, B., Pittman, K., Sutherland, R.J., & Whishaw, I.Q. (1982). Dissociation of the contributions of the prefrontal cortex and dorsomedical thalamic nucleus to spatially guided behavior in the rat. *Behavioural Brain Research*, *6*, 365–378.

Kolb, B., Sutherland, R.J., & Whishaw, I.Q. (1983). A comparison of the contribution of the frontal and parietal association cortex to spatial localization in rats. *Behavioral Neuroscience*, *97*, 13–27.

Landfield, P.W., Rose, G., Sandles, L., Wohlstader, T.C., & Lynch, G. (1977). Patterns of astroglial hypertrophy and neuronal degeneration in the hippocampus of aged, memory-deficient rats. *Journal of Gerontology*, *32*, 3–12.

Lansdell, H.C. (1968). Effect of extent of temporal lobe ablations on two lateralized deficits. *Physiology and Behavior*, *3*, 271–273.

Light, L.L., & Zelinski, E.M. (1983). Memory for spatial information in young and old adults. *Developmental Psychology*, *19*, 901–906.

Lippa, A.S., Pelham, R.W., Beer, B., Critchett, D.J., Dean, R.L., & Bartus, R.T. (1980). Brain cholinergic dysfunction and memory in aged rats. *Neurobiology of Aging*, *2*, 165–172.

Mahut, H. (1971). Spatial and object reversal learning in monkeys with partial temporal lobe ablations. *Neuropsychologia*, *9*, 409–424.

Mahut, H., & Cordeau, J.P. (1983). Spatial reversal deficit after amygdalo–hippocampal ablations. *Experimental Meurology*, *7*, 426–434.

Mahut, H., & Zola, S.M. (1973). A nonmodality specific impairment in spatial learning after fornix lesions in monkeys. *Neuropsychologia*, *11*, 255–269.

Mandler, J.M., Seegmiller, D., & Day, J. (1977). On the coding of spatial information. *Memory and Cognition*, *5*, 10–16.

Maki, W.S., Beatty, W.W., Hoffman, N., Bierley, R.A., & Clouse, B.A. (1984). Spatial memory over long retention intervals: Nonmemorial factors are not necessary for accurate performance on the radial maze by rats. *Behavioral and Neural Biology*, *41*, 1–6.

Maki, W.S., Brokofsky, S. & Berg, B. (1979). Spatial memory in rats: Resistance to retroactive interference. *Animal Learning and Behavior*, *7*, 25–30.

Markowitsch, H.J. (1982). Thalamic mediodorsal nucleus and memory: A critical evaluation of studies in animals and man. *Neuroscience and Biobehavioral Reviews*, *6*, 351–380.

Means, L.W., Hershey, A.E., Waterhouse, G.J., & Lane, C.J. (1975). Effects of dorsomedial thalamic lesions on spatial discrimination reversal in the rat. *Physiology and Behavior*, *14*, 725–729.

Milner, A.D., Ockleford, E.M., & Dewar, W. (1977). Visuo–spatial performance following posterior parietal and lateral frontal lesions in stumptail macaques. *Cortex*, *13*, 350–360.

Milner, B. (1965). Visually guided maze learning in man: Effect of bilateral hippocampal, bilateral frontal and unilateral cerebral lesions. *Neuropsychologia*, *3*, 317–338.

Milner, B. (1974). Hemispheric specialization: Scope and limits. In: F.O. Schmitt & F.G. Worden (Eds.), *The Neurosciences Third Study Program*. Cambridge, MA: MIT Press.

Mishkin, M. (1954). Visual discrimination performance following partial ablations of the temporal lobe: I. Ventral vs. lateral. *Journal of Comparative and Physiological Psychology*, *47*, 14–20.

Mishkin, M. (1978). Memory in monkeys severely impaired by combined but not by separate removal of amygdala and hippocampus. *Nature*, *273*, 297–298.

Mishkin, M., Vest, B., Waxler, M., & Rosvold, H. (1969). A reexamination of the effects of frontal lesions on object alternation. *Neuropsychologia*, *7*, 357–363.

Morris, R.G.M. (1981). Spatial localization does not require the presence of local cues. *Learning and Motivation*, *12*, 239–260.

Morris, R.G.M., Garrud, P., Rawlins, J.N.P., & O'Keefe, J. (1982). Place navigation impaired in rats with hippocampal lesions. *Nature*, *297*, 681–683.

Morrow, L., Ratcliff, G., Johnston, C. (1985). Externalising spatial knowledge in patients with right hemisphere lesions. *Cognitive Neuropsychology*, *2*, 265–273.

Murray, C.L., & Fibiger, H.C. (1985). Learning and memory deficits after lesions of the nucleus basalis magnocellularis: Reversal by physostigmine. *Neuroscience*, *14*, 1025–1032.

Newcombe, F. (1985). Neuropsychology *qua* interface. *Journal of Clinical and Experimental Neuropsychology*, *7*, 663–681.

Newcombe, F., & Russell, W.R. (1969). Dissociated visual perceptual and spatial deficits in focal lesions of the right hemisphere. *Journal of Neurology, Neurosurgery, and Psychiatry*, *32*, 73–81.

Niki, H. (1966). Response perseveration following the hippocampal ablation in the rat. *Japanese Psychological Research*, *8*, 1–9.

O'Keefe, J. (1976). Place units in the hippocampus of freely moving rats. *Experimental Neurology*, *51*, 78–109.

O'Keefe, J., & Nadel, L. (1978). *The hippocampus as a cognitive map*, Oxford: Clarendon Press.

Olton, D.S., Becker, J.T., & Handelman, G.E. (1979). Hippocampus, space and memory. *Behavioral and Brain Sciences*, *2*, 313–322.

Olton, D.S., & Collison, C. (1979). Intramaze cues and "odor trails" fail to direct choice behavior on an elevated maze. *Animal Learning and Behavior*, *7*, 221–223.

Olton, D.S., Collison, C., & Werz, M.H. (1977). Spatial memory and radial arm maze performance in rats. *Learning and Motivation*, *8*, 289–314.

Olton, D.S., & Feustle, W.A. (1981). Hippocampal function required for nonspatial working memory. *Experimental Brain Research*, *41*, 380–389.

Olton, D.S., & Papas, B.C. (1979). Spatial memory and hippocampal function. *Neuropsychologia*, *17*, 669–682.

Olton, D.S., Walker, J.A., & Gage, F.H. (1978). Hippocampal connections and spatial discrimination. *Brain Research*, *139*, 295–308.

Olton, D.S., Walker, J.A., & Wolf, W.A. (1982). A disconnection analysis of hippocampal function. *Brain Research*, *233*, 241–243.

Olton, D.S., & Wolf, W.A. (1981). Hippocampal seizures produce retrograde amnesia without a temporal gradient when they reset working memory. *Behavioral and Neural Biology*, *33*, 437–452.

Orbach, J., Milner, B., & Rasmussen, T. (1960). Learning and retention in monkeys after amygdala-hippocampus resection. *Archives of Neurology*, *3*, 230–251.

Oscar-Berman, M., & Zola-Morgan, S.M. (1980a). Comparative neuropsychology and Korsakoff's syndrome. I—Spatial and visual reversal learning. *Neuropsychologia*, *18*, 499–512.

Oscar-Berman, M., & Zola-Morgan, S.M. (1980b). Comparative neuropsychology and Korsakoff's syndrome. II—Two-choice visual discrimination learning. *Neuropsychologia*, *18*, 513–525.

Oscar-Berman, M., Zola-Morgan, S.M., Oberg, R.G.E., & Bonner, R.T. (1982). Comparative neuropsychology and Korsakoff's syndrome. III—Delayed response, delayed alternation and DRL performance. *Neuropsychologia*, *20*, 187–202.

Pandya, D.N., & Yeterian, E.H. (1984). Proposed neural circuitry for spatial memory in the primate brain. *Neuropsychologia*, *22*, 109–122.

Passingham, R.E. (1985). Memory of monkeys (*Macaca mulatta*) with lesions in prefrontal cortex. *Behavioral Neuroscience*, *99*, 3–21.

Penfield, W., & Milner, B. (1958). Memory defect produced by bilateral lesions in the hippocampal zone. *Archives of Neurology and Psychiatry*, *79*, 475–497.

Perlmutter, M., Metzger, R., Nezworski, T., & Miller, K. (1981). Spatial and temporal memory in 20 and 60 year olds. *Journal of Gerontology*, *36*, 59–65.

Peterson, C., & Gibson, G.E. (1983). Amelioration of age-related neurochemical and behavioral deficits by 3,4-diaminopyridine. *Neurobiology of Aging*, *4*, 25–30.

Petrides, M. (1985). Deficits on conditional associative-learning tasks after frontal and temporal lobe lesions in man. *Neuropsychologia*, *23*, 601–614.

Petrides, M., & Iversen, S.D. (1979). Restricted posterior parietal lesions in the rhesus monkey and performance on visuospatial tasks. *Brain Research*, *161*, 63–77.

Petrides, M., & Milner, B. (1982). Deficits on subject-ordered tasks after frontal- and temporal-lobe lesions in man. *Neuropsychologia*, *20*, 249–262.

Pezdek, K. (1983). Memory for items and their spatial locations by young and elderly adults. *Developmental Psychology*, *19*, 895–900.

Pohl, W. (1973). Dissociation of spatial discrimination deficits following frontal and parietal lesions in monkeys. *Journal of Comparative and Physiological Psychology*, *82*, 227–239.

Potegal, M. (1969). The role of the caudate nucleus in spatial orientation of rats. *Journal of Comparative and Physiological Psychology*, *69*, 765–764.

Potegal, M. (1971). A note on spatial–motor deficits in patients with Huntington's disease: A test of a hypothesis. *Neuropsychologia*, *9*, 233–235.

Potegal, M. (1972). The caudate nucleus egocentric localization system. *Acta Neurobiologia Experimentalis*, *32*, 479–494.

Pribram, K.H., Plotkin, H.C., Anderson, R.M., & Leong, D. (1977). Information sources in the delayed alternation task for normal and "frontal" monkeys. *Neuropsychologia*, *15*, 329–340.

Pribram, K.H., & Tubbs, W.E. (1967). Short-term memory, parsing and the primate frontal cortex, *Science*, *156*, 1765.

Ranck, J.B. Jr. (1973). Studies on single neurons in dorsal hippocampal formation and septum in unrestrained rats. *Experimental Neurology*, *41*, 461–555.

Ratcliff, G. (1979). Spatial thought mental rotation and the right cerebralhemisphere. *Neuropsychologia*, *17*, 49–54.

Ratcliff, G., & Newcombe, F. (1973). Spatial orientation in man: Effects of left, right and bilateral posterior cerebral lesions. *Journal of Neurology, Neurosurgery and Psychiatry*, *36*, 448–454.

Rosvold, H.E., & Szwarcbart, M. (1964). Neural structures involved in delayed-response performance. In J.M. Warren & K. Akert (Eds.). *The frontal granular cortex and behavior*. New York: McGraw Hill.

Salamone, J.D., Beart, P.M., Alpert, J.E., & Iversen, S.D. (1984). Impairment in T-maze alternation performance following nucleus basalis magnocellularis lesions in rats. *Behavioral Brain Research*, *13*, 63–70.

Samuels, I. (1972). Hippocampal lesions in the rat: Effects on spatial and visual habits. *Physiology and Behavior*, *8*, 1093–1098.

Samuels, I., Butters, N., & Fedio, P. (1972). Short term memory disorders following temporal lobe removals in humans. *Cortex*, *8*, 283–298.

Samuels, I., Butters, N., Fedio, P., & Cox, C. (1980). Deficits in short-term auditory memory for verbal material following right temporal lobe removals in humans. *International Journal of Neuroscience*, *11*, 101–108.

Schulman, S. (1964). Impaired delayed response from thalamic lesions. *Archives of Neurology*, *11*, 477–499.

Scoville, W.B., & Milner, B. (1957). Loss of recent memory after bilateral hippocampal lesions. *Journal of Neurology, Neurosurgery and Psychiatry*, *20*, 11–21.

Segal, M. (1982). Changes in neurotransmitter actions in aged rat hippocampus. *Neurobiology of Aging*, *3*, 121–124.

Semmes, J., Weinstein, S., Ghent, L., & Teuber, H.-L. (1963). Impaired orientation in personal and extrapersonal space. *Brain*, *86*, 747–772.

Smirni, P., Villardita, C., & Zappala, G. (1983). Influence of different paths on spatial memory performance in the block tapping test. *Journal of Clinical and Experimental Neuropsychology*, *5*, 355–360.

Smith, M.L., & Milner, B. (1981). The role of the right hippocampus in the recall of spatial location. *Neuropsychologia*, *19*, 781–793.

Squire, L.R. (1982). Comparison between forms of amnesia: Some deficits are unique to Korsakoff's syndrome. *Journal of Experimental Psychology: Learning, Memory and Cognition*, *8*, 560–571.

Stevens, R. (1981). Scopolamine impairs spatial maze performance. *Physiology and Behavior*, *27*, 385–386.

Stevens, R., & Cowey, A. (1972). Enhanced alternation learning in hippocampectomized rats by means of added cues. *Brain Research*, *46*, 1–22.

Stuss, D.T., & Benson, D.F. (1984). Neuropsychological studies of the frontal lobes. *Psychological Bulletin*, *95*, 3–28.

Sutherland, R.J., Kolb, B., Whishaw, I.Q., & Becker, J.B. (1982). Cortical noradrenaline depletion eliminates sparing of spatial learning after neonatal frontal cortex damage in the rat. *Neuroscience Letters*, *32*, 125–130.

Sutherland, R.J., Whishaw, I.Q., & Kolb, B. (1983). A behavioral analysis of spatial localization following electrolytic, kainate- or colchicine-induced damage to the hippocampal formation in the rat. *Behavioural Brain Research*, *7*, 133–153.

Sutherland, R.J., Whishaw, I.Q., & Regehr, J. (1982). Cholinergic receptor blockade impairs spatial localization by use of distal cues in the rat. *Journal of Comparative and Physiological Psychology*, *96*, 563–573.

Thomas, G.J., & Gash, D.M. (1985). Mammillothalamic tracts and representational memory. *Behavioral Neuroscience*, *99*, 621–630.

Thomas, G.J., & Spafford, P.S. (1984). Deficits for representational memory induced by septal and cortical lesions (singly and combined) in rats. *Behavioral Neuroscience*, *98*, 394–404.

Victor, M., Adams, R.D., & Collins, C.H. (1971). *The Wernicke–Korsakoff Syndrome*. Philadelphia: F.A. Davis Co.

Waddell, K.J., & Rogoff, B. (1981). Effect of contextual organization on spatial memory of middle-aged and older women. *Developmental Psychology*, *17*, 878–885.

Walker, J.A., & Olton, D.S. (1984). Fimbria–fornix lesions impair spatial working memory but not cognitive mapping. *Behavioral Neuroscience*, *98*, 226–242.

Wallace, J.E., Krauter, E.E., & Campbell, B.A. (1980). Animal models of declining memory in the aged: Short-term and spatial memory in the aged rat. *Journal of Gerontology*, *35*, 355–363.

Watt, J., Stevens, R., & Robinson, C. (1981). Effects of scopolamine on radial maze performance in rats. *Physiology and Behavior*, *26*, 845–851.

Waxler, M., & Rosvold, H.E. (1970). Delayed alternation in monkeys after removal of the hippocampus. *Neuropsychologia*, *8*, 137–146.

Whishaw, I.Q. (1985). Cholinergic receptor blockade in the rat impairs locale but not taxon strategies for place navigation in a swimming pool. *Behavioral Neuroscience*, *99*, 979–1005.

Whishaw, I.Q., & Kolb, B. (1984). Decortication abolishes place but not cue learning in rats. *Behavioral Brain Research*, *11*, 123–134.

Wilkinson, D.A. (1982). Examination of alcoholics by computed tomographic (CT) scans: A critical review. *Alcoholism*, *6*, 31–45.

Winocur, G. (1985). The hippocampus and thalamus: Their roles in short- and long-term memory and the effect of interference. *Behavioural Brain Research*, *16*, 135–152.

Wirsching, B.A., Beninger, R.J., Jhamandas, K., Boegman, R.J., & El-Defrawy, S.R. (1984). Differential effects of scopolamine on working and reference memory of rats in the radial maze. *Pharmacology, biochemistry and behavior*, *20*, 659–662.

Zola-Morgan, S., & Squire, L.R. (1985a). Medial temporal lesions in monkeys impair memory on a variety of tasks sensitive to human amnesia. *Behavioral Neuroscience*, *99*, 22–34.

Zola-Morgan, S., & Squire, L.R. (1985b). Amnesia in monkeys after lesions of the mediodorsal nucleus of the thalamus. *Annals of Neurology*, *17*, 558–564.

4
Humor and the Right Hemisphere: A Narrative Perspective

AMY M. BIHRLE, HIRAM H. BROWNELL, and HOWARD GARDNER

Humor is so taken for granted in most societies that its importance becomes evident only under unusual conditions—for example, when it is impaired or when it is altogether missing. The absence of a sense of humor can have debilitating social consequences and can deprive an individual of a significant source of pleasure.

The sense of humor after brain damage is of interest for both scientific and clinical reasons. On the scientific level, it is possible to discover aspects of the structure and functioning of humor through an investigation of the details of its decomposition. Clinically, the sense of humor after brain damage is important because an impairment may profoundly affect a patient's interaction with others, and possibly his or her return to an acceptable daily existence.

Though impairments in a sense of humor can be observed and studied following unilateral brain damage, a special methodological problem attends the study of a subset of humor, verbal humor in left-hemisphere-damaged (LHD) patients. Specifically, the ability to appreciate a verbal joke rests on normal language comprehension, which is often disrupted by aphasia. Hence, on a verbal humor task, it would be difficult to tease apart language- from humor-specific conceptual problems. For this reason, only a small portion of the work to date has focused on aphasic patients.

Patients with right-hemisphere damage (RHD) are interesting in this context because, unlike aphasic patients, they have largely preserved syntactic and semantic capabilities, and thus allow a ready analysis of humor comprehension. Such patients show striking impairments on humor comprehension tasks that cannot be attributed to a primarily linguistic impairment.

Another motivation for examining RHD patients' humor deficits stems from the clinical picture that these patients often present (Tucker, 1981). Their speech tends to be marked by frequent injection of personalizations, insignificant details, and tangential remarks. Often, these patients are considered excessively jocular, unconcerned, and inappropriate (Weinstein & Kahn, 1955). One goal of studying the humor comprehension of RHD patients is to clarify the striking clinical picture we ordinarily encounter following RHD.

Preliminary to an exploration of the humor deficit in patients with brain damage, it is appropriate to examine the models describing the normal appreciation of humorous stimuli. There are two major theories of humor to which most modern researchers subscribe: The incongruity and the incongruity-resolution theories.

Nerhardt (1977) favors the incongruity theory. He has argued that incongruity alone is sufficient to produce humor. Specifically, Nerhardt and others contend that humor based solely on surprise exists—if the surprise occurs in a nonthreatening environment. According to this model, the greater the discrepancy between an expectation and a punchline (in effect, the greater the surprise), the funnier the joke. However, this theory seems inadequate to account for many more complex forms of humor, such as those examined in studies reviewed below.

The incongruity-resolution theory (cf. Shultz, 1972; Suls, 1983) accounts for more complex forms of humor. When presented with a joke, the listener must formulate an expectancy based on the beginning of the joke, recognize the incongruity of the punchline, and then find a way to make the incongruous element fit with the body of the joke (Suls, 1983). This theory goes beyond the simpler incongruity model: equally at issue is the way in which a listener responds to the incongruity expressed in a joke. The incongruity is typically resolved by reinterpreting information in the joke or by invoking outside, real-world knowledge (i.e., knowledge of what normally happens in specific situations). Often, the incongruity can be removed by recognizing an ambiguous element in the body of the joke. In the case of puns, the ambiguity may be lexical, phonological, or syntactic in nature (for discussion, see McGhee, 1979).

It is important to emphasize the cognitive demands of humor. Incongruity can only be detected and resolved if the subject is able to recognize a surprising element and then figure out the ways in which that surprising element can be reconciled with one (or occasionally) more of the elements in the body of the joke. Supporting this analysis, McGhee (1979) argued that the pleasure we derive from humor stems from mastering the incongruity. This view is further substantiated by a finding by Zigler, Levine, and Gould (1966) of a curvilinear relationship between the level of difficulty and the funniness of a joke. Appreciation of a joke is low when the joke is either too easy or too difficult to understand.

A number of issues concerning right hemisphere disease and humor flow from discussion of normal humor comprehension. Specifically, one may examine the role of structural features in characterizing humor as a narrative genre. One can ask whether or not RHD patients are sensitive to the form of humor as suggested by the incongruity-resolution framework. Using the incongruity-resolution model, one can also ask at what stage in the humor comprehension process might RHD patients fail. One can look at the influence of content (e.g., the affective nature of the topic of a joke) on patients' appreciation of humor. Another issue is whether or not humor production is related to the ability to understand humor, and how RHD patients' humor production relates to their humor comprehension. A final issue concerns how the humor comprehension of RHD patients compares to that of other groups. We will consider the literature bearing on these issues and also suggest directions for future research.

One general comment is in order concerning the discussion of humor and of humor deficits presented in this review. We treat humor as a narrative genre requiring certain cognitive abilities. Though we recognize the importance of the affective elements in appreciation of a joke or cartoon, our focus (consistent with much of the literature) falls on the cognitive aspects of humor appreciation. Accordingly, a narrative framework serves as the basis for our account for why and how RHD patients fail at humor comprehension. We view the affective abnormalities of RHD patients as more relevant to their efforts at humor production.

Sensitivity to the Form of Humorous Stimuli

According to the aforementioned incongruity-resolution model of humor processing, a listener must first recognize a joke as a special genre before the comprehension process begins. The evidence thus far argues for relatively preserved sensitivity to joke form in patients with unilateral right hemisphere damage.

To determine whether RHD patients have a sensitivity to the form of a joke, it was necessary to devise a paradigm in which patients respond to stimulus items that are either formally appropriate or inappropriate to jokes as a genre. This can be done by devising different types of multiple choice endings for bodies of jokes. Specifically, surprise or incongruity is a necessary, formal element of humor as defined by various humor models. If patients are sensitive to the form, they should choose endings that are surprising with respect to the body of the joke. If patients are insensitive to the form, they may at times choose unsurprising, straightforward conclusions to the body of the joke.

This approach was used in a recent study by Brownell, Michel, Powelson, and Gardner (1983). The task required a subject to listen to the body of a joke and then to select a correct punchline from a set of four alternatives. The four alternatives included an affectively neutral ending that provided an unsurprising continuation of the joke's premises; an analogous but sad ending; a non-sequitur ending; and the correct, funny punchline. The non-sequitur ending, like the correct punchline, included an element of surprise—it did not follow directly from the joke's beginning. However, unlike the correct ending, the non-sequitur could not be coherently integrated with the premises on a second level to form an acceptable resolution to the joke's story. Thus, the choice of a non-sequitur indicated a preserved sensitivity to the surprise component of humor but an inability to integrate the body of the joke and its punchline into a satisfying interpretation. Results of this study indicated that RHD patients were more attracted to the non-sequitur endings than were the normal controls. Indeed, only the non-sequitur type endings clearly distinguished normal controls from patients with brain damage. The interpretation offered for this result is that while these patients have difficulty resolving incongruity, they appreciate that surprise is an essential component of humor.

The Brownell et al. (1983) study did not, however, provide a conclusive demonstration that RHD patients have an intact appreciation of the surprise element of jokes. Intact appreciation of the surprise element of narratives means more than simply being able to apply surprise; it means knowing when it is appropriate and when it is not. Because RHD patients were only tested on short story jokes (a narrative form that requires surprise), it is difficult to assess if RHD patients would demonstrate the same regard for form in other narrative genres.

More recent work by Bihrle, Brownell, Powelson, and Gardner (1986) demonstrates that RHD patients appreciate the formal differences between humorous and nonhumorous narrative stimuli. RHD patients and LHD aphasic patients were tested on an augmented pictorial version of the Brownell et al. (1983) study that included both a joke and a story condition. By using similar types of foils in both conditions, it was possible to examine more closely patients' appreciation of when surprise was a relevant feature of a stimulus and when it was not.

Alternative endings in both conditions varied with respect to the degree of surprise and coherence. All endings maintained the same artistic style and included the same main character as the initial frames. In the joke condition, alternative endings included the correct, funny punchline that was coherent and surprising, a straightforward, unsurprising resolution to the cartoon, and three types of non-sequitur endings. The non-sequitur endings themselves varied with respect to the degree of surprise, or, conversely, in how related they were to the theme of the first three frames.

An associated non-sequitur—the least surprising of the non-sequiturs—was related to an element of the story but did not provide a satisfying ending consonant with the main theme. For example, in an item that featured water, the associated non-sequitur also depicted water. A neutral non-sequitur had no particular relationship to the preceding three frames except that the same cartoon character and style of drawing was used. A humorous non-sequitur contained humorous, usually slapstick content that was unrelated to the content of the first three frames. For example, one humorous non-sequitur depicted a man slipping on a banana peel. The neutral and humorous non-sequiturs were equally surprising. Alternative endings in the story condition include the correct, straightforward ending and the three types of non-sequiturs. In the joke condition, patients were asked to choose the cartoon frame that "ended the comic strip in a funny way," whereas, in the story condition they were asked to "choose the frame that ends the story in an ordinary, not funny way."

In general, RHD patients performed more poorly than did the aphasic patients on both the joke and story conditions. Of more interest is the qualitative difference in patterns of errors produced by the two groups. Patients with RHD were most attracted to the non-sequitur endings in the joke condition—in particular the humorous non-sequiturs. In contrast, when they erred, aphasic patients typically chose the straightforward endings and they were least attracted to the humorous non-sequitur endings. In summary, RHD patients were most attracted to surprising, incoherent endings, whereas the aphasic patients were attracted to coherent, unsurprising endings. In the story condition, when they erred, both patient groups chose the associative non-sequiturs—endings that are low on the surprise continuum. The finding that RHD patients strongly preferred the humorous non-sequiturs in the joke condition, and seemed to avoid them in the story condition, indicates that these patients understand the formal differences between a joke and a story. Their attraction to the associated non-sequitur (the least surprising alternative) in the story condition, and not in the joke condition, provides further support for this claim. Given these results, one may speculate that left hemisphere structures suffice for the recognition of different narrative genres, but that right hemisphere structures are needed if the particular relationship between a punchline and the body of a joke is to be appreciated.

Additional, albeit indirect, evidence that RHD patients are sensitive to the form of humorous stimuli has been reported by Gardner, Brownell, Wapner, and Michelow (1983). This study examined subjects' interpretations of several types of potentially humorous narrative units:

Puns: What happened to the girl who swallowed a spoon?
She couldn't stir.

Tricks: Why do birds fly South for the winter?
It's too far to walk.

Puzzles: What speaks in every language in the world but never went to school?
An echo.

and **Foils:** Why do clouds move across the sky?
The wind pushes them.

In a test of RHD patients' appreciation of types of these potentially humorous narrative units, patients and normal controls were asked to rate puns, tricks, puzzles, and foils on the basis of funniness.

There was a large main effect of type: puns were rated as funniest; and foils the least funny. Most striking, however, was that the profile of rated funniness across types of items was the same for RHD patients as for the normal controls. In contrast, in another study looking at just one type of joke, RHD patients' funniness ratings were different than those produced by normal controls and aphasic patients (Gardner, Ling, Flamm, & Silverman, 1975). The degree of parallel performance reported here suggests that RHD patients are sensitive to distinctions among classes of items when these classes are defined at some metalinguistic level. There are of course several possible interpretations for this pattern. Perhaps patients and normal controls were relying on different strategies to rate funniness, with normal controls processing each item and RHD patients relying more on structural components and past memories. Finally, it must be conceded that the separate roles of content as well as form are not easily distinguished in this paradigm; in order to properly rate funniness, one must attend in part to context and not exclusively to form.

Analysis of the Processes Required for Humor Comprehension

Upon hearing the body of a joke, a listener must formulate expectancies (what might plausibly happen next), then recognize the violation of these expectancies in the punchline, and finally resolve the incongruity by forging an interpretation that links the punchline to the body of the joke. The evidence to date suggests that RHD patients are able to formulate appropriate expectancies and detect incongruity; it is their manipulation of an incongruous element that is abnormal.

Detection of Incongruity

McGhee (1979) stressed that a clear understanding of the normal course of events is essential for the enjoyment of humor, which plays on the incongruous depiction of these events. These is a growing body of evidence that RHD patients have a largely preserved sense of real-world knowledge. Recent work by Roman, Brownell, Potter, Seibold, and Gardner (1987) suggests that RHD patients, including patients sharing characteristic deficits on narrative humor tasks, have preserved script knowledge. The patients tested were typically able to list, in one form or another, the elements comprising well-rehearsed, familiar routines such as changing a flat tire or eating in a restaurant. Furthermore, when presented with two steps of a script, RHD patients could give the next step. Finally, they were able to say whether or not an element was part of a particular script. So, for example, RHD patients appreciated that "checking produce for freshness" is not part of eating in a restaurant.

Other support for RHD patients' appreciation of incongruity comes from their responses to noncanonical elements in stories. RHD patients seem to appreciate the incongruous nature of bizarre story elements even though they (RHD patients) do not react normally to the incongruity. In a study of narrative recall conducted by Wapner, Hamby, and Gardner (1981), normal controls typically omitted, regularized, or challenged noncanonical story elements. Patients with RHD recalled incongruous elements, but they characteristically felt the need to justify the noncanonical elements—even by confabulating a link. For example, in one story about a fishing trip, a father tells his son never to bring a flashlight on a boat after having just been rescued because of the flashlight. A RHD patient acknowledged that this was nonsensical, but rationalized the noncanonical element by saying, "He was trying to collect his son's insurance."

Resolution of Incongruity

Evidence from two humor studies (Bihrle et al., 1986; Brownell et al., 1983) suggests that RHD patients have difficulty resolving incongruity. They are particularly poor at establishing a coherent link between a punchline and the body of a joke as indicated by their confusion between the non-sequitur and the correct endings. An example helps to convey the nature of the difficulty. In the Bihrle et al. (1986) study, a patient was presented with the body of the joke and asked to choose the best punchline from several alternatives.

A man walks up to a lady in a crowded square. "Excuse me," he says, "have you seen a policeman anywhere around here?"

"I'm sorry," she said, "I haven't seen a sign of one."

Then the man said,

Straightforward Ending: "Damn, I've been looking for a half hour and can't find one".

Associated Non-Sequitur Ending: "My uncle is a cop."

Neutral Non-Sequitur Ending: "Baseball is my favorite sport."

Humorous Non-Sequitur Ending: "All of the wheels fell off my car."

Funny Correct Ending: "Then hurry up and give me your watch and pocketbook."

Upon hearing the body of the joke, a normal listener and an RHD patient would, very likely, formulate the same expectancy. Specifically, in the preceding example, a listener is led to believe that the man is in need of a policeman. The normal listener is quite surprised upon hearing the punchline. He resolves the incongruity by inferring that the man is a mugger, thus revising his interpretation of the body of the joke. More specifically, the listener reinterprets the intent behind the man's question concerning the whereabouts of a policeman. The apparent incongruity is resolved and humor results. The RHD patient is unwilling or unable to revise his initial interpretation upon hearing the punchline and, consequently, the punchline remains incongruous with respect to the body of the joke. One RHD patient acknowledged the incongruous nature of the punchline by commenting, "That's (i.e., the punchline) stupid." He then went on to confabulate an explanation for how the punchline should be interpreted: "He (the man) wants her watch now to see how long he's been waiting for a cop. He doesn't know what time it is." It is worth noting that because the patient did not revise his initial interpretation, in this case that the man is seeking a policeman, the punchline was an unsatisfying conclusion. Consequently, the patient chose the humorous non-sequitur ending, "All of the wheels fell off my car," because, "At least that's funny." In the final analysis, the RHD patient fulfills the surprise requirement of humor but grossly fails at the incongruity resolution stage of processing.

The experimental studies cited above, along with clinical observations, signal a problem resolving incongruity. This problem is not surprising in view of the fact that RHD patients have difficulty establishing coherence for simple narratives (Delis, Wapner, Moses, & Gardner, 1983; Huber & Gleber, 1982). Clearly, there are a number of contributing factors to RHD patients' aberrant performance at the incongruity resolution stage. We will review three of these: rigidity of interpretation, literalness, and inattention to relevant detail.

Rigidity of Interpretation

Rigidity of interpretation—an inability or unwillingness to revise an initial interpretation—may in part account for problems in resolving incongruity. The best evidence for this form of deficit comes from a study of inferential reasoning by Brownell, Potter, Bihrle, and Gardner (1986). In this study, RHD patients and normal controls were presented with two sentences and asked to treat them as one story. A sentence considered in isolation led to one interpretation ("incorrect inference") while two sentences considered together led to a different interpretation ("correct inference"). A representative stimulus item was as follows: "Sally took a pen and paper with her to meet the movie star. The article was to include famous peoples' opinions on nuclear power." After hearing the first sentence, a listener infers that Sally is seeking an autograph (the incorrect inference). However, after hearing the second sentence, the listener revises his or her interpretation and infers that Sally is instead a newspaper reporter (the correct inference) rather than that she is a fan seeking an autograph. Most important, the position (first or second) of the sentence supporting the incorrect inference was varied. The pattern of errors made by RHD patients suggests that getting stuck on a first association interfered with making the appropriate conceptual link between the two sentences. Specifically, RHD patients were most likely to answer incorrectly when the misleading information was presented in the first rather than in the second sentence.

The same rigidity may occur when RHD patients are presented with a joke. As the incongruity-resolution model predicts, upon hearing the body of a joke, a listener formulates a certain interpretation, and then generates a different interpretation upon hearing the punchline. Patients with RHD seem to lack the ability, or equally likely, the inclination to revise their initial interpretation in light of new information. Consequently, as is the case of the policeman joke cited earlier, the punchline is never reinterpreted to fit with the body of the joke and the intended humor is missed altogether. The RHD patient discussed was never able to make the correct inference—that the man is a mugger—inasmuch as he remained stuck on an initial interpretation.

It is worth noting that RHD patients may exhibit an inference deficit apart from a rigidity of interpretation (Joanette, Goulet, Ska, & Nespoulous, 1986). An inability to properly infer a "bridge" (i.e., a conceptual link) between a sentence in a discourse and what has come before (cf. Clark, 1977; Haviland & Clark, 1974) would greatly impair humor comprehension. It remains for future work to disentangle the respective contribution to poor humor comprehension of a rigidity of interpretation and a "bridging" deficit.

Literalness

Patients with right hemisphere disease tend to be literal. A strong preference for literal interpretations may adversely affect humor comprehension because incongruity resolution often requires viewing a word or phrase from two different perspectives, or focusing on different aspects of the meaning of a word or phrase. This tendency toward literalness has been documented from the single word level to the sentence level.

One recent study (Brownell, Potter, Michelow, & Gardner, 1984) suggests that RHD patients have a preserved sensitivity to denotation (dictionary meaning) and a selective insensitivity to connotation (e.g., metaphoric meaning). Unilaterally brain-damaged patients were presented with triads of words such as *warm–loving–cold* and asked to group together the two words that were most closely associated. Triads could be grouped on the basis of literal, denotative meaning relations (e.g., warm and cold) or metaphoric, more connotative similarity (e.g., warm and loving). Patients with RHD typically grouped on the basis of denotative meaning, while aphasic patients as a group focused on the connotative meanings. A subsequent study (Brownell, Simpson, Bihrle, Potter, & Gardner, submitted) compared RHD and LHD patients' sensitivity to alternative interpretations of two classes of words: adjectives (e.g., warm) with metaphoric alternative meanings and nouns with nonmetaphoric alternative meanings (e.g., cabinet: cupboard, advisors). Relative to the LHD patients, the RHD patients showed greater sensitivity to the nonmetaphoric alternative meanings of the adjectives used. Thus, the deficit does not apply equally to all alternative meanings of words, but seems to apply especially to connotatively similar metaphoric reinterpretations of select words. Similar results have been reported by Gardner and Denes

(1973) for visual stimuli. RHD patients again showed a preserved sensitivity to the denotations and a disturbed recognition of the connections of words.

Patients with RHD often focus on literal interpretations of phrases as well. On a test for comprehension of phrasal metaphor (Winner & Gardner, 1977), RHD patients were attracted to the literal depictions of metaphoric phrases while aphasics and normal controls more often chose the abstract depictions. For example, given the sentence "He was a broken man," RHD patients were apt to choose a picture of a man who had shattered into little pieces, while aphasics were more likely to choose a picture of a man crying. It is noteworthy that the aphasics usually reacted to the silliness of the literal depiction, while the RHD patients did not.

Finally, RHD patients' inclination to focus on literal interpretations extends to the sentential level. Concreteness at this level of language has implications for all types of verbal jokes. Recently, a number of investigators (Foldi, 1987; Hirst, LeDoux, & Stein, 1984; Myers & Linebaugh, 1981; Tompkins & Mateer, 1985) have reported that RHD patients tend to process indirect speech acts literally. Foldi (1987), for example, examined indirect requests which, by their nature, can elicit literal responses (i.e., responses that address the explicitly stated content of an utterance) or pragmatic, nonliteral responses (i.e., responses that address a speaker's intended meaning.)

As an illustration, consider the following vignette taken from Foldi's study: A boss is standing next to a secretary and is looking at a letter. A narrator's voice says, "The letter has many mistakes in it. The boss is displeased. The boss says, 'Is this the final copy of the letter?'" The RHD patients were markedly more literal than either normal control subjects or aphasic patients. When asked to choose the correct, appropriate response, the RHD patients showed a strong preference for the literal response: "Yes, it is" instead of the pragmatically correct response: "Let me do that again." In many cases, humor can only be appreciated if a listener is sensitive to the pragmatic, nonliteral aspect of language; yet this is precisely where RHD patients have difficulty.

Details

In order to resolve the incongruity inherent in a punchline, one must be able to identify the important details presented in the body of the joke. Normal listeners probably build a mental model of a joke (a tentative schema of the events described in the body of the joke that can be revised in the light of new and unexpected information) as they go along and encounter information that is inconsistent with the model (cf., Johnson-Laird, 1983). The documented insensitivity of RHD patients to relevant details in a narrative may contribute to their problems in resolving the incongruity of a joke. For example, it has been observed clinically that RHD patients have difficulty figuring out which detail matters. They have trouble organizing details appropriately; that is, they do not seem to appreciate which details are important and which are insignificant (Wapner et al., 1981). Using cartoon stimuli, Gardner et al. (1975) found that RHD patients had difficulty focusing on relevant details in a visually complex array—a problem that they speculated was conceptual rather than perceptual. RHD patients may simply not be able to identify the particular details that need to be revised to make sense of the joke.

In this line of research the role of a deficit in focusing on details remains to be determined. On some tasks, such as the inferential reasoning paradigm of Brownell et al. (1986), for example, an inability to focus on details did not appear to be a significant factor affecting RHD patients' performance. Yet, RHD patients may in fact have difficulty determining which of the many pieces of information in the body of a joke are relevant to the punchline. This hypothesis could be tested by giving patients jokes and highlighting the information relevant to fully understanding the joke. Alternatively, the hypothesis could be tested by giving patients "stripped down" versions of the joke and determining whether they have an easier time in understanding the intended humor.

The Roles of Joke Content and Affect in Humor Comprehension

Humorous material is traditionally considered affect-laden: Its purpose is most often to elicit an emotional response such as laughter. There is strong evidence that patients with right hemisphere lesions have particular difficulty with perception, appreciation, and expression of affective information (e.g., Heilman, Bowers, & Valenstein, 1985). For example, RHD patients have difficulty comprehending (Heilman, Bowers, Speedie, & Coslett, 1984) and expressing (Tucker, Watson, & Heilman, 1977) different emotional tones of voice. It is important to note that the study by Heilman et al. (1984) demonstrated an affective deficit over and above a more general prosodic disturbance disrupting comprehension of nonaffective information. Parallel difficulties have been noted in the nonlinguistic domain. For example, RHD patients often demonstrate an impaired ability to process (Bowers, Bauer, Coslett, & Heilman, 1985) and produce (Buck, & Duffy, 1980) emotional faces.

Initially, it was thought that RHD patients' well-documented difficulties with affective information would disrupt their appreciation of humor. Yet, a number of studies using strictly verbal materials have failed to show greater problems with emotionally laden content than with other types. It should be noted, however, that some investigators have found that RHD patients have greater difficulty recalling emotional than nonemotional material (Wechsler, 1973). Nonetheless, several recent studies fail to show a difference between affective and neutral linguistic material.

First, in a study conducted by Michelow, Brownell, Masson, Wapner, and Gardner (1982), subjects (aphasic patients, normal controls, and RHD patients) were asked to rank order sets of single words along dimensions such as "happiness" and "sadness." The RHD patients tested did not show greater difficulty with emotional than with nonemotional dimensions.

Similarly, in a test for comprehension of familiar phrasal metaphors, Winner and Gardner (1977) used both emotional (e.g., "heavy heart") and nonemotional (e.g., "loud tie") items. Again, there was no apparent difference in RHD patients' performance with the two types of metaphor.

At the level of single words, Brownell et al. (submitted) observed no differences in how likely RHD patients were to respond to the alternative meanings of more (e.g., "warm": loving) versus less (e.g., "quick": smart) affectively-laden metaphoric adjectives. Work by Etcoff (1984; 1985) also suggests that RHD patients have largely preserved abilities to process emotional words.

Wapner et al. (1981) presented patients with stories manipulated in systematic ways. Initially, it was hypothesized that stories that contained emotional elements would be more difficult for RHD patients to understand. In fact, RHD patients did not have more trouble with the emotional than with the nonemotional elements of their stories.

Finally, in a further effort to determine if RHD patients confuse emotions, Brownell et al. (1983) looked at how these patients treated sad endings in a joke completion task. If these patients had trouble distinguishing emotions, they might have chosen sad endings instead of the correct, funny punchlines. They showed no evidence of confusing the appropriately toned endings (i.e., the correct punchline) with inappropriately toned endings.

Perhaps the best evidence for viewing a humor comprehension deficit as a narrative rather than as an emotional deficit comes from a study recently conducted by Bihrle et al. (1986), in which a nonverbal cartoon completion task was administered to patients. RHD patients did almost as poorly in the story condition—an affectively neutral condition—as in the joke condition. Although this last set of studies does not prove that there is not a special problem with emotionally laden linguistic materials, it seems increasingly likely that whatever problems RHD patients have with the emotional realm cannot in themselves account for the observed humor deficits.

The Relation Between Humor Comprehension and Humor Production

Although, to our knowledge, the literature contains no formal studies of humor production in RHD patients, we have informally observed that they are quite fond of telling jokes. Often these jokes are inappropriate to the setting, and frequently the intended humor is lost on the audience. For example, we have noticed that these patients often tell off-color jokes and make fun of their spouses in the presence of hospital staff (e.g., "When she sits around the house—she really sits *around* the house"). The evidence just reviewed suggests that the affective nature of humor cannot, by itself, explain RHD patients' humor comprehension deficit. However, we speculate that the affective disturbance following right hemisphere damage may have a more significant impact on humor production. Perhaps RHD patients' failure to adequately assess a social situation leads to their inappropriate comments. They are unable to use their audiences' emotional reactions as a guide to modifying their behavior; nor are they sensitive to indirect cues such as speech intonation and facial expression. It may be their very inappropriateness and insensitivity that makes RHD patients' remarks funny; in effect, the surprise component contributes to the humor.

In one part of a study currently underway in our laboratory, we asked subjects to recount one of their own jokes. Some RHD patients refused to tell a joke. Others, though, were able to tell a well-formed short-story joke when asked to do so. Though surprising at first glance in light of their comprehension deficits, this ability may reflect no more than recitation of overlearned materials much like the Pledge of Allegiance. Furthermore, when RHD patients tell a joke with feeling and panache, it suggests that they have retained a performance, not merely some lines. We examined RHD patients' production deficit from another perspective by asking them to process novel material. Our results to date indicate that RHD patients as a group also show impairments when asked to retell jokes presented by someone else. Their immediate retellings often omit or garble the punchline, or omit critical elements in the body of the joke needed to "set up" the listener for the punchline.

The extent to which, and the reasons why, RHD patients are impaired in humor production will of course require further investigation. However, three preliminary conclusions have emerged. First, there is an affective deficit that limits patients' abilities to assess the social appropriateness of comments and attempts at humor; this deficit appears linked to a lack of empathy and inhibition. Second, there is a cognitive–narrative deficit that disrupts RHD patients' acquisition and retelling of new jokes and that is perhaps related to the comprehension disorders discussed earlier. Third, when a RHD patient does succeed at telling a short-story joke, this success most likely represents use of old knowledge that was acquired premorbidly and that is, therefore, relatively unaffected by the cognitive sequelae of brain pathology.

Comparison of Right Hemisphere-Damaged Patients' Performances with Those of Other Populations

An important issue concerns localization of functions relevant to humor comprehension. The results of a cartoon completion task (Bihrle et al., 1986) suggest that humor deficits result from damage specifically to the right hemisphere rather than from a general, nonlocalizable impairment. In addition to showing different errors patterns, aphasic LHD patients performed better than the RHD groups; of the 17 LHD patients tested, 7 showed a ceiling effect. This finding is not surprising in view of the fact that aphasic patients do relatively well on tests related to the incongruity-resolution process. They are also able to produce coherent, well-structured narratives (Ulatowska, North, & Macaluso-Haynes, 1981). In addition, LHD patients do not show the characteristic right hemisphere pattern on tests of literalness at the single word, phrase, or discourse levels (Brownell et al., 1984; Foldi, 1987; Winner & Gardner, 1977). The evidence suggests, then, that while LHD patients are by no means normal, their humor comprehension is relatively preserved.

A related issue concerns whether the problems comprehending humor discussed above are better viewed as "frontal" deficits that can occur from frontal damage to either hemisphere rather than as specifically right hemisphere deficits. Although preliminary, our data on this issue provide no support for an account for these deficits based on frontal lobe damage due to stroke, irrespective of side. Damage to part of the frontal lobe of the left hemisphere by itself does not seem sufficient for a humor deficit. On the Bihrle et al. cartoon-completion task, LHD patients with anterior damage in the vicinity of Broca's area performed better overall than the RHD group, and comparably to those with posterior damage. In addition, patients with anterior LHD did not make the same types of errors as did the RHD patients. Furthermore, a separation of the RHD patients in this study into subgroups with anterior or posterior damage revealed that both subgroups showed the characteristic appreciation of surprise but not coherence components of humor. Thus, humor deficits are not limited to patients with right anterior pathology.

Two cautionary notes are in order concerning these tentative conclusions. First, the work to date has included only small numbers of stroke patients in each of the relevant subgroups. Second, this work has examined the effects of damage to only some parts of the frontal lobe. For example, no systematic study of patients with exclusively right or left prefrontal brain damage has been conducted. Thus, the contribution to humor comprehension of these and other restricted portions of the frontal lobes is as yet unknown.

Future work must also explore the humor comprehension of brain-damaged patients with other etiologies—including dementing patients and closed head injury patients. In addition, it will be particularly important to determine how patients with psychiatric illnesses, especially affective disorders such as depression, understand humor. Perhaps humor comprehension can be preserved in the face of diminished appreciation. That is, some types of patients may understand what discourse units are jokes and why they are jokes. These same patients, though, may simply no longer enjoy jokes as they did premorbidly.

The RHD patients' comprehension of humor can also be compared to that of non-brain-damaged groups at different stages in the developmental sequence such as children. In general, RHD patients' comprehension of humor seems most similar to that of young children. Some researchers (Shultz, 1974; Shultz & Horibe, 1974) have reported that young children (first graders) are only sensitive to incongruity-based humor. Thus, when presented with riddles in which the punchline is modified to preserve either incongruity or resolution, children found the incongruity versions (i.e., with resolution removed) as funny as the original riddles. Like patients with right hemisphere disease, the children were sensitive to the surprise component of humor and insensitive to coherence. As shown in the example reprinted here from Shultz (1974), children found the original versions funnier than the resolution-removed and the incongruity-removed versions:

Why did the cookie crumble?
Original: Because its mother was a wafer so long.
Resolution removed: Because its mother was a wafer.
Incongruity removed: Because it was left in the oven too long.

Young children and RHD patients may fail to understand incongruity-resolution humor for some of the same reasons. Both groups seem to have difficulty viewing ambiguous words and phrases from two alternative perspectives. Shultz and Pilon (1973) presented ambiguous sentences to children of different ages and asked them to paraphrase the sentence and to point to pictures depicting the two possible meanings of each sentence. The first graders in this study were not sensitive to phonologic or syntactic ambiguity, which would explain why children at this age are unable to understand jokes which play on multiple meanings of a word or phrase.

Conclusion

On the theoretical side, RHD patients' performance on humor comprehension tasks demonstrates a clear dissociation between two aspects of humor: incongruity and resolution. Within the humor comprehension process, there are some abilities that are relatively spared and others that are quite disrupted. Patients with RHD seem to have an intact appreciation of the form of humorous stimuli. Converging evidence suggests that they have retained the ability to detect the incongruity inherent in humor. However, RHD patients demonstrate particular difficulty in resolving the incongruity of humor and integrating the elements into a coherent whole. We have suggested three major reasons for this failure at incongruity-resolution: rigidity of interpretation, literalness, and inattention to relevant detail.

Although research has documented that RHD patients fail at humor comprehension in certain characteristic ways, only indirect evidence suggests why they fail. Future research must test the candidate explanations for failure more directly. This may be accomplished in part by modifying jokes in systematic ways and ascertaining under which conditions RHD patients successfully comprehend a joke.

Although the emotional nature of humor is not sufficient to explain a humor deficit, future work must delineate how RHD patients' affective disorder affects humor production and comprehension. It has been suggested that an affective problem contributes to their sometimes aberrant humor production. Though it has been observed that RHD patients are quite fond of telling jokes, often at socially inappropriate times, this observation deserves careful systematic study.

On the clinical side, RHD patients' inability to understand humor helps explain why they interact inappropriately with others. The inability to understand complex units of language such as humor may extend to other types of speech acts (e.g., sarcasm, hyperbole) that are also vital to effective communication in natural settings. Successful management of this patient population must in part rest on an awareness of these and related deficits and how they may disrupt everyday social interaction.

Acknowledgments. This research was supported by grants NINCDS NS 11408, 06209, the Medical Research Service of the Veterans Administration, and Harvard Project Zero. We thank our colleagues Nancy Lefkowitz at Spaulding Rehabilitation Hospital and Anna Pomfret at the New England Rehabilitation Hospital for their cooperation in the research described in this chapter.

References

Bihrle, A.M., Brownell, H.H., Powelson, J., & Gardner, H. (1986). Comprehension of humorous and non-humorous materials by left and right brain-damaged patients. *Brain and Cognition*, *5*, 399–412.

Bowers, D., Bauer, R.M., Coslett, H.B., & Heilman, K.M. (1985). Processing of faces by patients with unilateral lesions. I. Dissociation between judgments of facial affect and facial identity. *Brain and Cognition*, *4*, 258–272.

Brownell, H.H., Michel, D., Powelson, J., & Gardner, H. (1983). Surprise but not coherence: Sensitivity to verbal humor in right hemisphere patients. *Brain and Language*, *18*, 20–27.

Brownell, H.H., Potter, H.H., Bihrle, A.M., & Gardner, H. (1986). Inference deficits in right brain-damaged patients. *Brain and Language*, *27*, 310–321.

Brownell, H.H., Potter, H.H., Michelow, D., & Gardner, H. (1984). Sensitivity to lexical denotation and connotation in brain-damaged patients. A double dissociation? *Brain and Language*, *22*, 253–265.

Brownell, H.H., Simpson, T.L., Bihrle, A.M., Potter, H.H., & Gardner, H. (submitted). Appreciation of metaphoric alternative word meanings by left and right brain-damaged patients.

Buck, R., & Duffy, R.J. (1980). Nonverbal communication of affect in brain damaged patients. *Cortex, 16*, 351–362.

Clark, H.H. (1977). Bridging. In P.N. Johnson-Laird & P.C. Wason (Eds.), *Thinking: Readings in Cognitive Science* (pp. 410–420). Cambridge: Cambridge University Press.

Delis, D.C., Wapner, W., Moses, J.A., & Gardner, H. (1983). The contribution of the right hemisphere to the organization of paragraphs. *Cortex, 19*, 43–50.

Etcoff, N.L. (1984). Perceptual and conceptual organization of facial emotions: hemispheric differences. *Brain and Cognition, 3*, 385–412.

Etcoff, N.L. (1985). The neuropsychology of emotional expression. *Massachusetts Institute of Technology Center for Cognitive Sciences Occasional Paper #31.*

Foldi, N.S. (1987). Appreciation of pragmatic interpretations of indirect commands: A comparison of right and left hemisphere brain-damaged patients. *Brain and Language, 31*, 88–108.

Gardner, H., Brownell, H.H., Wapner, W., & Michelow, D. (1983). Missing the point: The role of the right hemisphere in the processing of complex linguistic materials. In E. Perecman (Ed.), *Cognitive Processes in the Right Hemisphere* (pp. 169–191). New York: Academic Press.

Gardner, H., & Denes, G. (1973). Connotative judgements by aphasic patients on a pictorial adaptation of the semantic differential. *Cortex, 9*, 183–196.

Gardner, H., Ling, K., Flamm, L., & Silverman, J. (1975). Comprehension and appreciation of humor in brain-damaged patients. *Brain, 93*, 399–412.

Haviland, S.E., & Clark, H.H. (1974). What's new? Acquiring new information as a process in comprehension. *Journal of Verbal Learning and Verbal Behavior, 13*, 512–521.

Heilman, K.M., Bowers, D., Speedie, L., & Coslett, H.B. (1984). Comprehension of affective and nonaffective prosody. *Neurology, 34*, 917–921.

Heilman, K.M., Bowers, D., & Valenstein, E. (1985). Emotional disorders associated with neurological diseases. In K.M. Heilman & E. Valenstein (Eds.), *Clinical Neuropsychology* (pp. 377–402). New York: Oxford University Press.

Hirst, W., LeDoux, J., & Stein, S. (1984). Constraints on the processing of indirect speech acts: Evidence from aphasiology. *Brain and Language, 23*, 26–33.

Huber, W., & Gleber, J. (1982). Linguistic and nonlinguistic processing of narratives in aphasia. *Brain and Language, 16*, 1–18.

Joanette, Y., Goulet, P., Ska, B., & Nespoulous, J.-L. (1986). Informative content of narrative discourse in right brain-damaged right-handers. *Brain and Language, 29*, 81–105.

Johnson-Laird, P.N. (1983). *Mental models: Towards a Cognitive Science of Language, Inference, and Consciousness.* Cambridge: Harvard University Press.

McGhee, P.E. (1979). *Humors: Its Origin and Development.* San Francisco: W.H. Freeman and Company.

McGhee, P.E. (1983). Humor development: Toward a life span approach. In P.E. McGhee & J.H. Goldstein (Eds.), *Handbook of Humor research. Vol. 1: Basic issues* (pp. 109–134). New York: Springer-Verlag.

Michelow, D., Brownell, H., Masson, L., Wapner, W., & Gardner, H. (1982). On the sequencing of emotional and non-emotional verbal materials by organic patients. *Harvard Project Zero Technical Report*, No. 26.

Myers, P.S., & Linebaugh, C.W. (1981). Comprehension of idiomatic expressions by right-hemisphere-damaged adults. In R.H. Brookshire (Ed.), *Clinical Aphasiology Conference Proceedings: 1979* (pp. 254–261). Minneapolis: BRK Publishers.

Nerhardt, G. (1977). Operationalization of incongruity in humor research: A critique and suggestions. In A.J. Chapman & H. Foot (Eds.), *It's a Funny Thing, Humour* (pp. 47–51). London: Pergamon Press.

Roman, M., Brownell, H.H., Potter, H., Seibold, M., & Gardner, H. (1987). Script knowledge in right hemisphere damaged patients. *Brain and Language*, *31*, 151–170.

Shultz, T.R. (1972). The role of incongruity and resolution in children's appreciation of cartoon humor. *Journal of Experimental Child Psychology*, *13*, 456–477.

Shultz, T.R. (1974). Development of the appreciation of riddles. *Child Development*, *45*, 100–105.

Shultz, T.R., & Horibe, F. (1974). Development of the appreciation of verbal jokes. *Developmental Psychology*, *10*, 13–20.

Shultz, T.R., & Pilon, R. (1973). Development of the ability to detect linguistic ambiguity. *Child Development*, *44*, 728–733.

Suls, J.M. (1983). Cognitive processes in human appreciation. In P.E. McGhee and J.H. Goldstein (Eds.), *Handbook of Humor Research. Vol. 1: Basic Issues* (pp. 39–57). New York: Springer-Verlag.

Tompkins, C.A., & Mateer, C.A. (1985). Right hemisphere appreciation of prosodic and linguistic indications of implicit attitude. *Brain and Language*, *24*, 185–203.

Tucker, D.M. (1981). Lateral brain function, emotion, and conceptualization. *Psychological Bulletin*, *89*, 19–96.

Tucker, D.M., Watson, R.T., & Heilman, K.M. (1977). Affective discrimination and evocation in patients with right parietal disease. *Neurology*, *27*, 947–950.

Ulatowska, H.K., North, A.J., & Macaluso-Haynes, S. (1981). Production of narrative and procedural discourse in aphasia. *Brain and Language*, *13*, 345–371.

Wapner, W., Hamby, S., & Gardner, H. (1981). The role of the right hemisphere in the apprehension of complex linguistic mateials. *Brain and Language*, *14*, 15–32.

Wechsler, A.F. (1973). The effect of organic brain disease on recall of emotionally charged versus neutral narrative texts. *Neurology*, *23*, 130–135.

Weinstein, E.A., & Kahn, R.C. (1955). *Denial of Illness, Symbolic and Physiological Aspects*. Springfield, IL: Charles C. Thomas.

Winner, E., & Gardner, H. (1977). The comprehension of metaphor in brain-damaged patients. *Brain*, *100*, 717–729.

Zigler, E., Levine, J., & Gould, L. (1966). Cognitive processes in the development of children's appreciation of humor. *Child Development*, *37*, 507–518.

5
Nonverbal Cognitive Disturbances in Aphasia

GUIDO GAINOTTI

The relationship between verbal and nonverbal cognitive disorders of aphasic patients is a controversial and difficult problem: controversial in the relevance of the problems it, in turn, raises; difficult because empirical data obtained in aphasic patients do not lend themselves to clear and univocal interpretations. From the standpoint of theory, the study of nonverbal cognitive disturbances of aphasic patients could help to clarify (in well-controlled conditions) one of the questions that has arisen during centuries of sharp but inconclusive debates between phylosophers and psychologists; namely, the relationship between language and thought. From the practical standpoint, however, the authors who have approached the problem with either clinical or experimental procedures have often been surprised by the contrast between conclusions reached when considering aphasics as a whole, and those reached from a more analytic consideration of the data. When considered as a group, patients with aphasia generally show significant nonverbal cognitive impairment; however marked variability can be observed both between subjects and within subjects (i.e., the same patient can obtain poor results on certain nonverbal cognitive tasks, but perform satisfactorily on other ones).

Three different approaches can be used to clarify this issue: (*a*) a review of the history of aphasia in the literature; (*b*) a description of the data gathered by different authors in more recent investigations; (*c*) seeing how these data fit with more general theoretical interpretations.

History of Research on the Relations Between Language and Cognition in Aphasia

Early Proponents of a Nonverbal Cognitive Impairment in Aphasia

In 1863 while stressing the need for better analysis of the complex functional disorder constituting "*lésion du langage articulé*," Laborde noted that Broca (1861) had perhaps too easily admitted that intelligence was intact in these patients. Upon superficial examination, these subjects appear to be intelligent because they tend to make up for their inability to speak with very intense mimicking expressions. If asked to overcome this deficit by writing out their oral expressions, however, their written language is shown to be as impaired as their oral language.

This line of thought was developed by Trousseau (1865), who questioned the ability of aphasic patients to express their thoughts by gestures and by drawings, though admitting that intellectual impairment was far from being homogeneous in aphasia. As a proof he reported the case of a patient ("Paquet") who, though completely unable to express himself orally or by writing, was still perfectly able to play droughts and dominoes. Trousseau argued that aphasic patients may suffer from a special kind of amnesia that leaves intact certain aspects of memory, instead selectively impairing memory for words, necessary for the expression of thoughts. According to Trousseau, certain forms of intellectual activity that can be performed without the ancillary use of words, may be therefore spared in aphasic patients, but the great majority are impaired because the highest cognitive activities cannot be performed without those thought formulas ("formules de la pensée") that are represented by words.

Some years later, the existence of a cognitive impairment in aphasia largely exceeding the linguistic sphere was confirmed by Finkelnburg (1870), who explained it from a different theoretical point of view. He reported five patients with aphasia who, in addition to a language disorder, showed a number of nonverbal receptive and expressive symbolic disturbances: (*a*) the inability of a violinist to write down and to recognize musical notes; (*b*) the incapacity of a businessman to recognize the value of various pieces of money; (*c*) the inability a Prussian embassy official to understand conventional manners, insignia of rank and social class; though a practicing and strict Catholic, he was also unable to decipher liturgical symbols performed by the priest during the mass. In addition to these disturbances, some of these patients also showed an inability to acquire and retain the meaning of new signs, gestural impairments, or an inability to comprehend pantomimes. Finkelnburg argued that the notion of aphasia as a purely verbal disorder is inadequate to explain these findings and proposed that aphasia may be considered as an aspect of "asymbolia," defined as a "partial or complete loss of the capacity to comprehend and express concepts by means of acquired signs." The work of Finkelnburg was so influential that both Kussmaul (1877) and Jackson (1879) accepted his proposal. Jackson (1879), in particular, repeatedly mentioned that disorders in pantomime production and recognition are often observed in aphasic patients and acknowledged that there is in aphasia "a loss or defect in symbolizing relations of things in any way."

Language and Cognitive Disturbances According to the Authors of the "Noetic School"

Some decades later, the concept of a cognitive impairment in aphasia concomitant to (and possibly underlying) the properly verbal disturbance was strongly argued by Marie (1906), Head (1926), and Goldstein (1948). Marie put the question in very sharp terms; that there is in aphasia a marked impairment of intelligence in general. He offered as proof the following examples of nonverbal cognitive disturbances: the inability to tell the time on a clock or to make simple arithmetical operations; the incapacity to recognize or to execute simple symbolic gestures; the loss of previous professional skills, such as writing down or recognizing musical notes in a musician or cooking an egg in a chef. These examples suggest that Marie did not use the term "intelligence" to denote the capacity to solve new problems by means of thought processes, but rather in a sense loosely corresponding to the present construct of "semantic memory," defined by Tulving (1972) as "our permanent knowledge about the language and basic facts of the world." This suggestion is supported by Marie's definition of intelligence, considered as a sphere "where are stored, among others, various aspects of didactically acquired knowledge."

Head (1926), too, reported clinical observations showing that the cognitive impairment of aphasic patients looms larger than the purely linguistic domain. One of his patients, was unable to draw spontaneously an elephant (though he had been familiar with these animals during a long stay in India); another was unable to make simple arithmetic operations, to set a clock, to judge the relative value of two pieces of money, to set a table for a meal, and so on. Head, however, was less sharp than Marie in evaluating the meaning of these patterns of behavior; he did not presume aphasia to be due to a general loss of intelligence. Pursuing the line of thought of Finkelnburg and of Jackson, Head considered them as the consequence of a defect in the purposeful use of symbols. To be sure, he maintained that the greatest intellectual impairment of aphasic patients is observed on those tasks that require an intermediate symbolic activity between the initial conception and the final execution of the act. Aphasic patients are, therefore, not impaired on cognitive tasks that can be performed with a simple perceptual activity, but are usually deficient when an intermediate (verbal or nonverbal) symbolic activity is requested by the nature of the task.

Goldstein (1948) was the author of the so-called "Noetic school" who had the greatest influence upon the further course of thought on the relationships between linguistic and cognitive disturbances in aphasia. Studying with Gelb the naming disorders of brain-injured veterans of the First World War (Gelb & Goldstein, 1924), he concluded that naming impairment is only a manifestation of a more basic cognitive disorder, defined as a loss of the abstract attitude. Goldstein (1948) claimed that two different attitudes can be normally adopted toward the world: (*a*) the concrete attitude, which is manifested by the tendency to stick to the immediate sensory aspects of a given situation, and (*b*) the abstract attitude, whereby the subject detaches himself from the immediate sensory components of a situation and relies upon abstract rules and general concepts.

When the abstract attitude is disrupted by brain damage, then words lose their capacity to "organize" the world into general categories and instead remain simple sound sequences associated to concrete, particular objects. In support of this interpretation, Goldstein (1948) reported a number of examples showing that aphasic patients sometimes use "pseudonames" (i.e., individual words associated to individual objects) instead of "names" used in their abstract sense to denote members of concepts or of categories. For example, a patient was unable to utter the word "knife" when presented with the corresponding object, but named it "a bread-knife" when it was presented with a piece of bread, "pencil sharpener" when it was associated with a pencil or an "apple parer" when the knife was shown with an apple. Goldstein maintained that for this patient objects are not members of general classes, but are unique and concrete and the name is the sound complex belonging to that object.

Aphasia, and in particular amnestic aphasia, is thus the result of a basic deficit of abstractive ability; the disruption of verbal behavior is only an effect of a more general disorder. A comparative analysis of the fine-grained positions of these clinical authors evinces two different lines of thought. On one hand, authors such as Finkelnburg (1870), Marie (1906), and Goldstein (1948) assume that language disorders must be considered as only an aspect of a more basic cognitive impairment. On the other hand, authors such as Trousseau (1864) and Head (1926) consider the nonverbal cognitive impairment as the consequence, and not the cause, of the properly verbal disturbance. To be sure, these authors maintain that the intellectual impairment of aphasic patients is a byproduct of their language disorder and attribute this defect to the weakness of thought processes, which cannot use in their development an adequate verbal (or even nonverbal) symbolic support.

First Psychometric Investigations Using "Verbal" and "Nonverbal" Intelligence Tests in Aphasia

A weak aspect of the position of noeticians was the frail body of data provided by these authors in support of their theories. Their arguments consisted of individual case studies showing an association between aphasia and nonverbal symbolic disturbances (Finkelnburg), loss of the abstract attitude (Goldstein), or loss of previously acquired knowledge and skills (Marie), but they did not demonstrate that a necessary relationship exists between language and cognitive disturbances.

In a well-controlled study by Weisenburg and McBride (1985), however, verbal and nonverbal cognitive tasks were assigned to unselected groups of patients with aphasia, normal subjects, and nonaphasic patients with damage to the right hemisphere of the brain. The results clearly showed that the relationship between aphasia and nonverbal cognitive impairment is not absolute. Some of their patients showed clear evidence of a non-verbal cognitive impairment resulting from (or in any case related to) their language disorder, but others continued to function at good cognitive levels, despite gross language disturbances (e.g., speech suppression, neologistic jargon). It must be admitted, however, that nonverbal intelligence tasks used by Weisenburg and McBride were not necessarily of the kind best suited to control for the assumptions of the noetic authors. For the most part these tests (including formboards, mosaics, picture interpretation, and drawing tests) involved more perceptual–motor and visual–spatial abilities than the symbolic or categorial activities considered by the authors of the "noetic" school as typically impaired in aphasic patients. Therefore, the safer inference that can be drawn from the results of Weisenburg and McBride is that aphasia does not result in a diffuse and homogeneous intellectual defect. This cautious position was, indeed, assumed by Weisenburg and McBride who rightly drew attention to some subtle relationships between language and intelligence both in normal subjects and in patients with aphasia. In particular, they stressed the existence of two different sources of variance accounting for results obtained on cognitive tasks by aphasic patients: (*a*) the nature of the task itself because, as already stressed by Trousseau (1864) and Head (1926), some tasks require the intermediate use of inner language whereas other tasks can usually be performed without recourse to a covert verbal formulation; and (*b*) the cognitive style of the individual and his thought habits, in as much as some persons consistently rely upon the intermediate use of language for various kinds of cognitive activities, whereas other persons make a greater use of visual images and hence are less dependent from verbal formulation in their mental activities. According to Weisenburg and McBride, this cognitive typology could explain why greater or smaller defects of general intelligence may appear in different aphasic patients.

More Recent Trends: Methods, Theories, and Lines of Research

The Application of Methods Drawn from the Experimental Psychology to the Study of Nonverbal Cognitive Disorders in Aphasia

More recent studies have investigated the question of the nonverbal cognitive disorders of aphasic patients with designs and procedures drawn from the field of experimental psychology. A systematic group of studies performed by Teuber and associates in patients with penetrating brain wounds seems particularly demonstrative of this approach. These authors not only tried to avoid possible selection bias (including in their research also severe aphasic patients), but also made a serious attempt to construct really nonverbal cognitive tasks, modeling to a large extent their procedures on tasks used to study problem-solving and discrimination-learning behaviors in subhuman primates. Results were generally in agreement with the view of an aphasic cognitive defect not directly referable to a language disorder. On a variety of tactual and visual complex discrimination tasks, patients with aphasia scored only slightly worse than patients with damage to the right-hemisphere (Semmes, Weinstein, Ghent, & Teuber, 1954; Weinstein, Teuber, Ghent, & Semmes, 1955), but aphasic patients scored much worse than any other control group on two tests: (*a*) the Gottschaldt "hidden figures" test, in which patients had to identify a geometrical figure embedded in the context of a more complex design (Teuber & Weinstein, 1956) and (*b*) a "conditional reaction" test, in which patients had to learn to choose a triangle rather than a circle when these figures appeared on an horizontally striped background, and the circle instead of the triangle, when the background stripes were vertical (Weinstein et al., 1955). Weinstein (1964) has commented that both these tests involve figure–ground relations, as in the former the figure is embedded in the background, whereas in the latter the background provides the critical cue for the selection of the figure. He has also expressed the opinion that these findings support the view that language disorders of aphasic patients are but an aspect of a more basic cognitive disorder, that can be regarded as a defective organization and selection of linguistic as well as nonlinguistic material.

Aphasia and Impairment of Conceptual Thinking According to Bay

The hypothesis of a relatively specific cognitive defect in aphasia, which could be ascribed neither to a diffuse intellectual impairment nor to a purely linguistic disorder, has been more recently renewed by Bay (1962, 1964, 1974). He claimed that aphasia must be considered as a loss of concepts that may be expressed in words and not as a specifically linguistic disorder. In support of his claim, Bay (1962, 1964) reported the observation that patients with aphasia often obtain very poor performances on modeling tasks. Thus, an aphasic patient did not reproduce the essential detail of the wheel while modeling a wheel-barrow and an other roughly reproduced a giraffe as an animal, but without including the characteristic features of a long neck and long and slender legs. Furthermore, to stress the analogy between verbal and nonverbal aphasic disturbances, Bay claimed that patients who are unable to name a missing feature in an incomplete drawing are also unable to complete it. In agreement with Goldstein, Bay made a clear distinction between the sensory–motor components of the aphasic symptom complex and the core of the aphasic disturbance (the "genuine aphasia") that he identified with the amnestic aphasia.

Although Bay and Goldstein concur in acknowledging that naming disturbances lie at the core of the aphasic disintegration and are therefore subsumed by a basic cognitive disorder, they disagree as for the nature of this preverbal cognitive defect. Goldstein (1948) assumes that anomia is due to a categorization impairment, resulting from the loss of the abstract attitude. Bay (1962, 1964), on the other hand, maintains that the same poor differentiation and actualization of concepts subsumes verbal and non-verbal symbolic disabilities.

The distinction proposed by Bay between poor differentiation and defective actualization of concepts seems exciting because it anticipates the present distinction between semantic representation and semantic retrieval disorders. It must be acknowledged, however, that Bay's observations are open to criticism from the methodological point of view. For example, Zangwill (1969) has noticed that Bay gives no control data concerning the modeling capacities of brain-damaged patients without aphasia. It is, therefore, impossible to say if poor modeling performances of aphasic patients are due to a poor "differentiation and actualization of concepts" (Bay, 1974) or due to the influence of associated variables such as, for example, constructional apraxia, which is independent from aphasia but often observed in aphasic patients with lesions located in the posterior temporoparietal areas of the left hemisphere (Benton, 1973).

Intelligence and "Associative" Disorders in Aphasia According to De Renzi and His School

Although theoretically interesting, the Bay's views had not been supported by a sufficient body of empirical data obtained with appropriate methodology. This gap was filled by De Renzi and others, in a series of well-controlled neuropsychological investigations (De Renzi & Faglioni, 1965; De Renzi, Faglioni, Savoiardo, & Vignolo, 1966; De Renzi, Faglioni, Scotti, & Spinnler, 1972; De Renzi, Pieczuro, & Vignolo, 1968; De Renzi, Scotti, & Spinnler, 1969). The cognitive tasks used in these important investigations can be subdivided into two partly separable subgroups:

1. Intelligence tests using nonverbal abstract material as stimuli and requiring the patients to analyze or to organize these stimuli to solve a visual–spatial or a logical problem. Examples of the intelligence tests are:
   A. The Raven Progressive Matrices (Raven, 1938), a well-known test of visual–spatial intelligence that has been used by Arrigoni and De Renzi (1964) De Renzi and Faglioni (1965), and Colonna and Faglioni (1966) in the standard version, and by Basso, De Renzi, Faglioni, Scotti, and Spinnler (1973) in the simpler, colored form (Raven, 1947).
   B. Elithorn Perceptual Maze test (Elithorn, 1955), requires patients to find a path through the maximum number of dots arranged at the intersection of lattices within a rectangular configuration. This test was used by Colonna and Faglioni (1966).
   C. Gottschaldt hidden figures test, has already been used by Teuber and

Weinstein (1956) in patients with penetrating missile wounds of the brain, and was used again in patients with spontaneous cerebral pathology by Russo and Vignolo (1967).

D. Weigl sorting test (Weigl, 1927), requires patients to classify a set of wooden figures according to different criteria, such as form, color, and size. A modified version of this test, using five different categorization criteria, was used by De Renzi, Faglioni, Savoiardo and Vignolo in 1966.

2. "Associative" tests using concrete, pictorial material as stimuli and requiring the patients to match two different features, belonging to the conceptual representation of the same object. Examples of associative tests are:

A. The "Coloring drawings of objects" test, in which patients are requested to associate the line drawings of simple objects (such as a cherry, cigar, or banana) with the corresponding characteristic color (De Renzi & Spinnler, 1967; De Renzi, Faglioni, Scotti, & Spinnler, 1972; Basso, Faglioni, & Spinner, 1976).

B. The "meaningful sounds recognition" test requires patients to associate a familiar nonverbal sound (such as the barking of a dog) with a simple picture that represents the source of the sound (Spinnler & Vignolo, 1966; Faglioni, Spinnler, & Vignolo, 1969).

C. The "use of objects" test (Ideational Apraxia), requires patients to show how to use such familiar objects as a hammer or a toothbrush (De Renzi, Pieczuro, & Vignolo, 1968).

Results obtained in such a complex and articulated series of investigations, can be summarized as follows:

1. Aphasic patients show consistent impairment on both the intelligence and the associative tests.
2. Only part of this impairment, however, seems specific; the remainder seems generically due to the effect of brain damage, being observed both in aphasic and in nonaphasic, brain-damaged patients.
3. If performances obtained by aphasic patients are compared with those of nonaphasic right-hemisphere-damaged patients (considered as the best control group in terms both of extent and of localization of brain injury), then the intellectual impairment of aphasic patients seems rather specifically linked to the abilities tapped by the associative tests and by the Weigl test of abstract thinking. This result seems fairly consistent with the views of those noetic authors who had considered the selective cognitive impairment of aphasic patients as consisting of either a conceptual disorder or a defect of abstract thinking.
4. No difference can be found, however, between patients with aphasia and those with right-hemisphere damage on the tests more highly loaded in visual–spatial factor (such as the Raven Matrices and the Elithorn maze test). This finding confirms that visual–spatial intelligence tests do not

necessarily rely upon language processing and can also be performed by using non–verbal, visual–perceptual strategies.
5. The correlations between severity of language impairment and results obtained on the nonverbal intelligence and associative tasks are rather variable as a function both of the parameter chosen to evaluate the severity of aphasia and of the type of cognitive task. In a few cases only, however, these correlations reach a striking level of statistical significance.

Although these data confirm the association between aphasia and nonverbal cognitive disturbances, De Renzi et al. did not maintain that their results supported the hypothesis of a functional relationship between linguistic and nonverbal cognitive disturbances. Stressing the fact that the correlations between the severity of aphasia and the results obtained on associative and intelligence tests were significant, but not absolute, they surmised that anatomically contiguous, but functionally independent structures might subsume language and high level nonverbal abilities.

The frequent association between aphasia and nonverbal cognitive disturbances would therefore be due to the anatomic contiguity between brain areas subserving language and cortical structures underlying nonverbal cognitive functions and not to the existence of an intrinsic linkage between these two orders of behavioral disturbances. Results obtained by De Renzi et al. have been replicated by investigations conducted with different procedures in various laboratories. For example, the rather selective impairment of aphasic patients on tasks requiring them to match nonverbal meaningful sounds with the corresponding pictures has been confirmed by others (Doehring, Dudley, & Coderre, 1967; Strohner, Cohen, Kelter, & Woll, 1978, Varney, 1980). Analogously, the inability of aphasic patients to match line drawings of objects with the corresponding colors has been confirmed by Assal and Buttet (1976) and Cohen and Kelter (1979).

On the other hand, archibald, Wepman, and Jones (1967), Van Dongen (1973) and Gainotti, D'Erme, Villa, and Caltagirone (1986) have found no significant difference in the performance of patients with aphasia and those with right-hemisphere damage on nonverbal tests of visual spatial intelligence, such as the Raven's Matrices. Thus, the existence of a specific pattern of nonverbal cognitive impairment in aphasia, with selective disruption of conceptual–associative and of categorical–abstractive capacities seems supported by a very large body of data. Results obtained studying impaired pantomime interpretations and other kinds of nonverbal representative disturbances in aphasia further strenghten this contention.

Pantomime Interpretation and Gestural Expression in Aphasia

As already mentioned, both Trousseau (1864) and Finkelnburg (1870) noticed that pantomime comprehension and gestural expression are often impaired in patients with aphasia. Jackson (1879), Marie (1906), and Head

(1926) also paid special attention to this phenomenon, which can be considered as one of the most consistent nonverbal cognitive disorders described by the authors of the noetic school.

It is, then, amazing that during a half century these observations were either generally neglected or at least did not prompt any systematic investigation of this phenomenon in patients with aphasia and with nonaphasic brain damage. This form of oblivion was, perhaps, due to the fact that most neurological schools of the time considered these observations as confusing and resulting from the fact that at the end of the last century, apraxia (i.e., a gestural disorder resulting from an inability either to conceive or to execute a certain pattern of movement) had not yet been isolated by Liepmann (1905). The observations of gestural recognition and expression disturbances in aphasic patients were, therefore, considered as due to the co-occurrence of aphasic and apraxic disorders and not, as the noetic authors had assumed, to a general disturbance in the capacity to express or comprehend symbols in any modality.

Goodglass and Kaplan (1963) were the first to test these alternative interpretations. They administered a battery of tests of pantomime and of gesture production to a group of patients with aphasia and another with nonaphasic brain damage. They reasoned that if a "central communication disorder" subsumes the verbal and gestural impairment of patients with aphasia, then a high level of correlation should exist between the severity of aphasia and the degree of gestural defect. If, on the contrary, language and gestural disorders are produced by a lesion of anatomically contiguous, but functionally independent areas of the left hemisphere, then no clear relationship should be found between these two orders of disturbances. Goodglass and Kaplan indeed found no significant correlation and concluded that defects in gesture production of aphasic patients are not part of a "central communication disorder," but are instead the consequence of an apraxic impairment independent from aphasia.

Very different conclusions were reached, however, in several independent studies that investigated the relationships between language and gestural impairment at the receptive (and not at the expressive) level. The studies by Gainotti and Ibba (1972), Pickett (1974), Duffy, Duffy, and Pearson (1975), and Gainotti and Lemmo (1976) consistently showed (*a*) that patients with aphasia perform significantly worse than any other group with nonaphasic brain damage on nonverbal tests of pantomime recognition; (*b*) that a deficit in pantomime identification is strongly related to the severity of language comprehension disorders. These authors, therefore, concluded that—at least at the receptive level—language and gestural disturbances are clearly associated in patients with brain damage. This contention has been subsequently confirmed by research conducted in many other laboratories (for reviews cf. Peterson & Kirshner, 1981; Christopoulou & Bonvillian, 1985) and has been further developed and probed by Duffy (Duffy & McEwen, 1978; Duffy & Duffy, 1981; Duffy, Duffy, & Mercaitis, 1984; Duffy & Watkins, 1984).

This author, who has explicitly renewed the conception of "asymbolia," prompting the translation into English of the main article of Finkelnburg (Duffy & Liles, 1979) has provided the following pieces of experimental evidence in support of his contention:

1. A strong relationship exists in aphasia between scores obtained on tests of pantomime expression and of pantomime recognition. Furthermore, both these measures are highly correlated with scores obtained on tests of verbal expression, verbal comprehension and communicative ability (Duffy & Duffy, 1981). These findings are in substantial agreement with those obtained by Gainotti and Ibba (1972), Gainotti and Lemmo (1976), and Netsu and Marquardt (1984).
2. The strength of the relationship between pantomimic impairment and limb apraxia is substantially reduced when the variance associated with aphasia is partialled out, whereas the strength of the relationship between pantomime deficit and aphasia is only marginally affected by eliminating the variance due to limb apraxia (Duffy & Duffy, 1981).
3. Strong similarities can be observed between the gestural and the speech characteristics of fluent and nonfluent aphasic patients. The former produce a variegated and abundant number of gestures, embedded in a flow of off-target and undifferentiated movement sequences; the latter are characterized by a paucity and simplicity of gestures and by a more critical attitude toward their own performance (Duffy et al., 1984). Similar results have been obtained by Cicone, Wapner, Foldi, Zurif, and Gardner (1979).
4. Manipulation of variables affecting performance on verbal comprehension tasks have similar effects on tests of pantomime recognition. In particular, the presence of semantically related response choices significantly reduces scores obtained both on a verbal recognition and on a pantomime recognition test (Duffy & Watkins, 1984). These results are consistent with those of Seron, Van Der Kaa, Remitz, and Van Der Linden (1979), Daniloff, Noll, Fristoe, and Lloyd (1982), and Varney and Benton (1982) who found that aphasic patients with defective pantomime recognition show a marked response bias in favor of semantically related foils.

According to Duffy and Duffy (1981), these data "do support the view that the clinical syndrome of aphasia typically consists of related problems in both verbal and nonverbal behaviors; . . . the occurrence of verbal and nonverbal deficits may be explained either as a consequence of a central symbolic deficit, or as a problem of verbal mediation of nonverbal behaviors."

Two main objections have been addressed to the Duffys' claim that a central symbolic defect might subsume both gestural and verbal disturbances of aphasic patients: The first is that defects in pantomime comprehension are shown by a percentage of aphasic patients, which never exceeds the two-thirds of them (Daniloff et al., 1982; Varney, 1980, 1982; Christopoulou & Bonvillian, 1985). This suggests that there is not a necessary relationship between verbal and gestural disturbances and that the conclusions regarding a central symbolic disorder are, therefore, premature.

The second objection is that pantomime recognition defects are more strongly related to reading comprehension than to auditory comprehension disorders (Varney, 1978, 1980, 1982; Varney & Vilensky, 1980). According to Varney, these results indicate that spoken language processing and pantomime recognition are somewhat separate functions, and that the determinants of impaired pantomime comprehension are more specific in nature than a central symbolic disorder. Both these objections challenge a strong version of the "central symbolic deficit theory" (considering in terms of symbolic disruption each component of the aphasic symptom-complex), but neither is necessarily at variance with a theory simply stating that a functional relationship exists between some aspects of verbal and nonverbal (pantomime recognition) disorders in aphasia.

In fact, that pantomime comprehension disorders are observed only in a subgroup of aphasic patients can be explained by assuming that aphasia is a complex, multicomponent syndrome and that only one (or some) of these components are intimately linked to pantomime recognition impairment. As for Varney's observations that pantomime recognition defects are more strongly related to reading comprehension than to auditory comprehension disorders, they do not exclude the hypothesis of a functional relationship between verbal and gestural comprehension disorders. They simply indicate that, in addition to a common general factor, which according to Duffy and Duffy (1981) accounts for the greatest part of variance of scores obtained on verbal and gestural tests, there are also specific modality factors, accounting for a minor part of the variability observed on the same tests. Varney's data are in good agreement with this view in as much as they show that a strong intertest relationship exists between pantomime comprehension, reading comprehension, and auditory comprehension scores, but that the correlations between the two visual tests (pantomime recognition and reading comprehension) are slightly higher than the correlations between each of them and the auditory verbal comprehension test. (Varney, 1978).

Other Forms of Nonverbal Symbolic or Representative Disturbances in Aphasia

Turning to other kinds of symbolic defects described in patients with aphasia, we will now discuss two recent studies, that explore nonverbal symbolic (or representative) disturbances other than pantomime recognition in aphasia. One of these investigations (Wapner & Gardner, 1981) tried to check the central symbolic deficit theory by administering to normal subjects, patients with aphasia and those with right-hemisphere brain damage seven different categories of visual symbols. These categories ranged from purely pictorial displays (such as pictures of objects) to purely linguistic symbols (such as written phrases) passing through intermediate hybrid forms, such as trafic signs or trademarks. Subjects were requested to choose the correctly depicted symbol from among a set of four alternatives, discarding the incorrect alternatives. In the case of pictured objects, incorrect depictions had additions, deletions, or misplaced elements; in written phrases, the incorrect stimuli were syntactically or semantically anomalous. The central symbolic deficit theory predicts that aphasic patients should exhibit an across-the-board difficulty in processing the whole gamut of visual symbols. Results challenged this expectation because purely linguistic stimuli posed special difficulties for aphasic patients only whereas purely pictorial stimuli posed comparable difficulties for patients with right-hemisphere damage.

The investigation by Gainotti, Silveri, Villa, and Caltagirone (1983) tested the same hypothesis by means of a different methodology: they studied the capacity of patients with aphasia and nonaphasic brain damage to draw from memory the form of objects having a characteristic shape. The study was prompted by the observations of Head (1926) and Bay (1962, 1964) that aphasic patients are often unable to incorporate characteristic features or essential details in their reproduction of objects. To overcome the objection by Zangwill (1969) that poor modeling performances of aphasic patients may be independent from aphasia and due to a concomitant constructional apraxia, the scores of the "drawing objects from memory" task were covariated for scores obtained on a copying-drawings task. In the "drawing from memory" task, patients with aphasia were found to be significantly more impaired than both normal controls and patients with right-hemisphere damage even when the influence of constructional apraxia was partialled out. In agreement with pantomime recognition findings, these data suggest that a functional relationship exists in aphasia between at least some components of the aphasic syndrome and various kinds of expressive and receptive nonverbal symbolic disturbances.

The disagreement with the results obtained by Wapner and Gardner (1981) can perhaps be understood if we look at the task demands of the picture recognition task used by these authors. Their test required more a careful discrimination of the formal aspects of the depicted object than a full appreciation of its meaning, whereas the main difficulty of aphasic patients seems to reside just in the incapacity to fully understand the meaning of a (verbal of nonverbal) symbolic referent and not in a poor discrimination of its formal attributes. This fact has been clearly demonstrated by De Renzi, Scotti, and Spinnler (1969), who administered two kinds of nonverbal pictorial tasks to patients with damage to the right and left (usually aphasic) hemispheres of the brain. Some of these tasks were particularly demanding from the perceptual point of view, whereas one (the object-figure identification test) required the patients to match real objects with pictures of the same objects. As figures were not identical to objects, but differed from them in terms of both form and color, this test required a correct identification of the meaning of the figures and could not be performed on the basis of a simple perceptual matching. A double dissociation was found between patients with damage to the right hemisphere (who were significantly more impaired on the perceptual tests) and those with left-hemispheric damage (who were selectively unable to perform the object–figure identification test). As choosing the correctly depicted pictorial symbols used by Wapner and Gardner (1981) required more a careful perceptual analysis than a full appreciation of the meaning of the depicted figures, the contrast with other investigations on nonverbal symbolic disturbances of aphasic patients can be explained.

Nonverbal Cognitive Impairment and Semantic–Lexical Disintegration in Aphasia

Data reviewed in previous sections of this chapter have shown that the correlations between severity of aphasia and results obtained on conceptual, symbolic, or cognitive tests are never so high as would be requested by a strong version of the "asymbolic theory" of aphasia. This fact led De Renzi et al. to advance the hypothesis that anatomically contiguous, but functionally independent structures subsume language and nonverbal cognitive functions. However, Gainotti et al. argued that the incomplete value of the correlations between cognitive impairment and severity of aphasia is hardly surprising if we consider the complexity of the aphasic disintegration and the high number of perceptual, motor, and properly linguistic components that intermingle in the various clinical forms of aphasia. They made the following assumptions: (*a*) only some components of aphasia are intimately linked with the nonverbal cognitive impairment; (*b*) these components should be "central" (and hence linguistic in nature); and (*c*) a selective relationship could exist between nonverbal cognitive impairment and disruption of the semantic–lexical level of integration of language.

Two main reasons support this last assumption. First, phonology and syntax, being related to form rather than to content, should be less relevant to cognitive functions than lexical–semantics, which can be considered as the level where the knowledge of the world interacts with the linguistic system. Second, if a common pathophysiologic mechanism subsumes verbal and nonverbal cognitive disturbances of aphasic patients, then a similar locus of impairment should be observed both in verbal and in nonverbal symbolic disturbances. Now we have seen at the end of the last section that the main impairment of aphasic patients on nonverbal symbolic tasks seems to lie at the deep (semantic) and not on the surface (perceptual) level of symbol processing: these patients are often unable to fully understand the meaning of a pictorial or of a gestural symbol, although they show a good discrimination of its formal attributes. It seems, therefore, logical to assume that the same patients should also be unable to fully appreciate the semantic value of a lexical unit, although being able to discriminate the sounds of language and to use the surface morphosyntactic structures.

The hypothesis of a close relationship between nonverbal cognitive disturbances and breakdown of the semantic–lexical systems was tested in a series of neuropsychological investigations conducted on large groups of unselected aphasic and nonaphasic patients with right- and left-hemispheric damage. Two main purposes oriented these studies:

1. The first was to control the specificity of the aphasics' cognitive impairment by comparing results obtained by aphasic patients and nonaphasic patients with right-hemispheric damage on non-verbal tasks of (*a*) symbolic comprehension, (*b*) symbolic expression, (*c*) conceptual thinking, (*d*) classificatory activities, (*e*) temporal sequencing ability, and (*f*) visual-spatial intelligence.
2. The second was to control the relationships between nonverbal cognitive impairment and semantic–lexical disturbances by comparing results obtained on the same tasks by aphasic patients clustered on the ground of two different parameters: (*a*) the overall severity of aphasia, (*b*) the presence or the absence of a semantic–lexical disintegration at the receptive level.

The presence of a semantic–lexical impairment was studied at the receptive level because at the expressive level other aphasic disturbances (such as a speech suppression, severe anomia, or phonemic jargon) can mask the semantic disorder and render its detection difficult. At the receptive level, on the contrary, semantic–lexical disturbances can be more easily detected by asking patients to discriminate the meaning of a given word, by choosing it from an array of semantically similar alternatives. The "verbal sound and meaning discrimination test" (Gainotti, Caltagirone, & Ibba, 1975), which has already proved to be very useful from this point of view, was administered to all aphasic patients, and the number of semantic errors obtained was used to separate aphasics with and without semantic–lexical disturbances. The quantity of communication possible in oral expression and in auditory comprehension was taken as an index of the severity of aphasia: mild, moderate, and severe. Table 5.1 reports the results obtained on the nonverbal cognitive tasks and the outcome of the comparisons between aphasic and nonaphasic (normal subjects or patients with right-hemispheric damage) patients and between aphasic patients with and without semantic-lexical impairment.

The former comparison was highly significant on (*a*) two receptive and expressive symbolic tasks (the symbolic gesture comprehension test and

TABLE 5.1. Results obtained on various cognitive tasks by normal controls, patients with aphasia and nonaphasic right hemisphere-damage. Scores obtained by aphasics with and without semantic–lexical disturbances are also separately reported.

Cognitive activity explored by the test	Test	Comparison between aphasic and nonaphasic patients					Comparison between aphasic patients with and without semantic disturbances		
		Controls	(*p*)	RHDP	(*p*)	AP	With SD	(*p*)	Without SD
Symbolic comprehension[a]	Symbolic gesture comprehension	9.96	(NS)	9.85	(< .01)	8.13	7.98	(< .05)	9.14
Symbolic expression[b]	Drawing from memory	33.78	(< .01)	28.08	(< .01)	21.59	17.52	(< .05)	26.33
Conceptual thinking[c]	Conceptual relationships[a]	1.03	(NS)	1.92	(< .01)	5.51	6.71	(< .01)	3.44
Classificatory activity[d]	Class inclusion[g]	2.36	(NS)	1.88	(NS)	4.48	6.22	(< .01)	2.00
	Class intersection[g]	2.00	(NS)	1.53	(< .01)	6.58	9.04	(< .01)	3.08
Temporal Sequencing[e]	Temporal rules induction[g]	10.60	(< .01)	16.20	(NS)	18.25	21.65	(< .01)	13.33
Visual–spatial intelligence[f]	Raven's coloured matrices	24.67	(< .01)	21.40	(NS)	20.53	19.45	(< .01)	24.67

Abbreviations: AP = patients with aphasia; NS = not significant; RHDP = patients with right-hemisphere damage; SD = semantic disturbance.

[a] Gainotti & Lemmo, 1976.

[b] Gainotti, Silveri, Villa, & Caltagirone, 1983.

[c] Gainotti, Miceli, & Caltagirone, 1979.

[d] Gainotti, Carlomagno, Craca, & Silveri, 1986.

[e] Gainotti, Nocentini, Sena, & Silveri, 1986.

[f] Gainotti, D'Erme, Villa, & Caltagirone, 1986.

[g] Higher scores indicate a greater cognitive impairment.

the drawing objects from memory test) already mentioned in previous sections of this chapter; (*b*) a test of conceptual thinking (the test of conceptual relationships); and (*c*) a task of classificatory activity (the class intersection test) requiring the patients to keep simultaneously in mind two different criteria of classification of concrete realistic pictures.

However, no significant difference was observed between aphasic patients and nonaphasic patients with right-hemispheric damage on (*a*) an easy task of class inclusion; (*b*) a well-known test of visual–spatial intelligence (the Raven Coloured Matrices) and (*c*) a sequencing ability test (the temporal rules induction test) requiring the patients to discover the structure underlying the repetition of binary events, such as the simple alternance between two events (i.e., A–B, A–B, A–B, etc.) or the asymmetric alternance between the same events (i.e., A–B–B, A–B–B, A–B–B, etc.).

The results of these tests are in substantial agreement with the view that nonverbal cognitive functions specifically impaired in aphasia concern (*a*) the capacity to use symbols to express concepts (and to understand concepts expressed by symbols); (*b*) the capacity to use knowledge stored in the conceptual representation of objects; and (*c*) the capacity to abstract single features of concepts and to use them for classificatory activities. As for the second aim of this group of investigations, the data reported in Table 5.1 consistently show that a strong relationship exists in aphasia between nonverbal cognitive impairment and semantic–lexical verbal disturbances. In all the cognitive tests taken into account, the worst results were consistently obtained by aphasic patients with semantic–lexical disturbances, whereas those free from impairment of the semantic structures were undistinguishable not only from nonaphasic patients with right-hemisphere damage, but sometimes even from normal subjects.

The difference between aphasic patients with and without semantic–lexical disturbances was not simply due to the greater severity of aphasia shown by the former, inasmuch as the relationships between cognitive impairment and overall severity of aphasia (not reported in Table 1) hardly reached the level of statistical significance. These findings clearly show that not all the components of the aphasic symptom-complex are equally related to the impairment observed on nonverbal cognitive tasks. Factors linked either to the processing of auditory verbal stimuli (such as phoneme discrimination or phoneme identification disorders) or to the emission of well-articulated word forms (such as a syndrome of phonetic disintegration) may disturb the process of verbal communication, but also be quite independent from the nonverbal conceptual or symbolic disturbance. Impairment on nonverbal cognitive tasks is, however, generally observed in connection with a disruption of the semantic–lexical structures. It is this fact that confirms the existence of an intimate link in aphasia between impairment on nonverbal conceptual or symbolic tasks and disruption of the semantic–lexical level of language integration.

Main Interpretations Advanced to Explain the Association Between Verbal and Nonverbal Cognitive Disturbances in Aphasia

Two main interpretations have been advanced to explain the association between verbal and nonverbal cognitive disturbances in aphasic patients. The first, which is favored by neurologically oriented authors, assumes the functional independence between verbal and nonverbal cognitive disorders and explains their co-occurrence as due to the anatomic contiguity between brain structures subserving language and nonverbal cognitive functions. The second interpretation, which is preferred by psychologically oriented authors, takes one of the following alternative positions: (*a*) the cognitive functions usually impaired in aphasia—though on first sight "nonverbal"—do require, in reality, the intermediate structuring role of inner language; (*b*) a basic cognitive disorder subsumes both the verbal and the nonverbal conceptual/symbolic disturbances of aphasic patients.

The "Anatomical Contiguity" Hypothesis

The main argument supporting the anatomical contiguity hypothesis consists in the lack of a complete correlation between severity of aphasia and scores obtained on the cognitive tasks. This argument has been raised both by Goodglass and Kaplan (1963) against the "central communication disorder hypothesis" and by De Renzi et al. in favor of the "anatomical contiguity hypothesis." However, this argument is weakened by the fact that not all the components of the aphasic symptom-complex are equally related to a nonverbal cognitive disturbance. Some components, more directly linked to the "peripheral" aspects of fluent articulation and of phonemic hearing processing, contribute to the severity of aphasia but are apparently not related to the nonverbal cognitive impairment. Other components, namely the more nuclear disturbances of the semantic–lexical structures, seem on the contrary intimately linked to the nonverbal conceptual/symbolic or categorical impairment.

Moreover, it has been observed that variables capable of affecting performance on verbal tasks have similar effects on nonverbal conceptual or symbolic tasks. For example, both on verbal comprehension tasks (Gainotti et al., 1975) and on nonverbal tasks of sound recognition (Spinnler & Vignolo, 1966) and of pantomime identification (Seron et al., 1979; Varney & Benton, 1982) aphasic patients show a marked response bias in favor of making semantically related errors. Analogously, Duffy and Watkins (1984) have shown that on both a verbal comprehension task and a pantomime identification task the number of errors made by aphasic patients increases when response choices include semantically related alternatives. These findings suggest that language comprehension and pantomime recognition are not functionally independent disturbances, but instead share a common pathophysiologic mechanism, consisting either in the inability to access the full semantic representation of a concept through a (verbal or nonverbal) symbol or in a degredation of the semantic/conceptual representation itself.

Finally the anatomical contiguity hypothesis has never been submitted to a rigorous empirical control by systematically searching for clear dissociations between linguistic and cognitive disturbances, which should exist according to this hypothesis. In particular, it has not been convincingly shown that clear cut nonverbal cognitive disturbances can be observed in patients with left-hemispheric damage but without aphasia, although such cases would obviously bear a stronger support to the theory.

To be sure, some cases reported in the neuropsychological literature document the sparing of certain artistic abilities in patients with severe aphasia (Alajouanine, 1948; Basso & Capitani, 1985) and these cases could be considered evidence of the functional independence between linguistic and cognitive disturbances in aphasia, It must be noted, however, that the artistic abilities spared in these cases concerned either pictorial expression or musical capacities (i.e., functions that may be predominantly mediated by the right hemisphere). These cases, therefore, although certainly intriguing, cannot counter-balance the evidence suggesting a functional interconnection between verbal and nonverbal disorders in aphasia.

Much more difficult is, in our opinion, to choose between the two interpretations advanced to explain the functional relationship between language impairment and nonverbal cognitive disturbances. The first assumes that the intermediate use of language helps in general to solve apparently nonverbal tasks and that, accordingly, language disturbances also impair performances obtained on these tasks. On the other hand the second interpretation maintains, that disruption of a deep basic capacity subsumes both language disturbances and nonverbal cognitive disorders of aphasic patients. The first of these interpretations is perhaps the simplest one as it views a partly unexpected observation (such as the nonverbal cognitive impairment of aphasic patients) as a byproduct of the obvious linguistic defect of these subjects. The fact that aphasics fail even on tasks considered as genuinely nonverbal as they were originally devised to study discrimination learning and problem solving in nonhuman primates; (Teuber & Weinstein, 1956) can be explained by assuming with Vygotsky (1962) that in humans thinking has become so deeply restructured by language that language disorders usually hamper every form of thought processes.

The "Verbal Mediation" Hypothesis

Some data in favor of the "verbal mediation hypothesis" can be found in studies conducted in patients suffering either from a visual–verbal or from a tacto–verbal disconnection syndrome, which showed a rebound of misnaming on nonverbal tasks performances (Geschwind & Fusillo, 1966; Lhermitte & Beauvois, 1973; Beauvois, Saillant, Meininger, Lhermitte, 1978; Piras, De Vreese, Ortu, D'Onofrio, and Agnetti, 1985). Geschwind and Fusillo (1966) demonstrated this phenomenon on a color sorting task in which they gave the patient a chip telling him that the chip was "red" and that he had to sort all the red chips. By specifying that the color to be sorted was red, they forced the patient to solve an apparently nonverbal task by means of a verbal strategy. Consequently when he misnamed as red one shade of green, he put it with the red chips by saying "this is another shade of red" and then picked up all the similar shades of green.

Lhermitte and Beauvois (1973), on the other hand, showed that misnaming can influence results obtained on a drawing from memory task. In one case of misnaming, the patient drew what he had said ("a bird") instead of what he had seen (a mouse), and in other instances he included in his drawings details belonging in part to the stimulus and in part to the misnamed object.

The influence of misnaming on nonverbal pantomimic activity has also been advocated by Beauvois et al. (1978) in a case of tactual aphasia and was more explicitely demonstrated by Piras et al. (1985) in a case of optic aphasia. The patient described by Beauvois et al. (1978) was selectively unable to name tactually presented objects, which he recognized, as suggested by his ability to mime their use. Occasionally, however, he showed semantic "parapraxias" (e.g., he used a toothbrush like a comb). Suspecting that these miming errors might be due to an implicit semantic paraphasia, Beauvois et al. (1978) inhibited verbal mediation by giving special instructions and by placing adhesive tape on the patient's mouth. Under these conditions, semantic parapraxias no longer appeared and the patient flawlessly mimed the use of tactually presented objects.

Even more clear was the disturbing influence of misnaming on nonverbal tasks in the case of optic aphasia reported by Piras et al. (1985). This patient failed to name a visually presented bow, but began to mime its use correctly when requested by the examiner. In the meantime, however, he misnamed the bow as "a racket" and this wrong verbalization induced a shift in his miming activity, which changed from bending the bow to smashing a ball. The same phenomenon was observed on a coloring drawing task, when the patient colored in red the line drawing of a tangerine, after having misnamed it as a "slice of watermellon."

A finding that strongly supports the generalization to the nonverbal behavior of aphasic patients of observations made in patients with modality-specific naming disorders is the similarity between the types of naming errors usually made by these two groups of patients. Several authors (Lhermitte & Beauvois, 1973; Beauvois et al., 1978; Gil, Pluchon, Toullat, Micheneau, Rogez, & Lefevre, 1985) have shown that the naming errors made by patients with modality-specific aphasia are, in general, semantically related to the target.

On the other hand, Gainotti et al. (1975) have shown that semantic paraphasias constitute the most characteristic type of naming error made by aphasic patients with semantic–lexical receptive disturbances (i.e. by patients who show the highest incidence of nonverbal cognitive disturbances). It is therefore possible that these errors may have a similar influence on nonverbal task performance both in patients with modality-specific aphasia and in aphasic patients with multimodal semantic–lexical disturbances. It is also possible that the marked response bias in favor of semantically related errors shown by aphasic patients on both meaningful sound recognition tasks (Spinnler & Vignolo, 1966; Faglioni et al., 1969) and pantomime recognition tasks (Seron et al., 1979; Varney & Benton, 1982; Duffy & Watkins, 1984) may be due to their tendency to produce semantic paraphasias.

The "Basic Preverbal Defect" Hypothesis

Many authors have questioned the idea that the aphasic patient's nonverbal cognitive impairment is simply a byproduct of language disruption, and have advocated that both verbal and nonverbal cognitive impairment of these patients may be subserved by a basic preverbal defect. No general agreement exists, however, about the nature of this preverbal impairment, and all the hypotheses proposed so far are not very convincing: they either explain only certain aspects of aphasic cognitive defect, or they are not supported by a sufficient body of empirical data. The following theories are examined below: (*a*) the "central symbolic deficit" theory; (*b*) the "loss of the abstract attitude" hypothesis; (*c*) the "defective identification of individual features" hypothesis; and (*d*) the "semantic/conceptual representation disruption" hypothesis.

The "Central Symbolic Deficit" Hypothesis

The central symbolic deficit hypothesis has already been illustrated, both in reporting the seminal observations of Finkelnburg (1870) and in describing the recent developments of this theory by Duffy et al. As the main arguments favoring this theory have already been exposed in sufficient detail, we will focus here our attention on the objections that have been addressed to the Duffy's claims and that generally arise from persons interested in language rehabilitation. The bulk of these objections consists in the fact that some severe aphasic patients who fail to reacquire adequate spoken language skills after intensive and prolonged speech therapy may, nevertheless, acquire aspects of a manual or visual communication system (for a review, see Christopoulou & Bonvillian, 1985). Glass, Gazzaniga, and Premack (1973) have, for example retrained patients with global aphasia who had failed to reacquire spoken language through conventional speech therapy by using a visual procedure adapted from the work of Premack (1971) with chimpanzee. The results of their training program (based on learning nonverbal visual symbols) have been encouraging because some of their patients made considerable gains in the capacity to communicate basic needs to their families. These data are not consistent with the central symbolic deficit hypothesis because according to this theory, damage to central symbol processing mechanisms should prevent aphasic patients from learning visual or gestural symbolic systems. It must be acknowledged, however, that the outcomes of therapeutic interventions using these procedures are extremely variable. Inasmuch as a detailed account of the linguistic functions impaired and spared in these patients is usually not reported, results obtained up to date cannot resolve the question whether a central symbolic defect is present or not in aphasia.

The "Loss of the Abstract Attitude" Hypothesis

The loss of the abstract attitude hypothesis will be only shortly discussed here because some of the data supporting this theory have already been reported in the section devoted to the work of Goldstein. We will only recall here that Goldstein et al. developed several nonverbal tests of sorting, classification, and the like for assessing disorders of the abstract attitude in patients with brain-damage (Goldstein & Scheerer, 1941) but that results obtained with these tests on unselected groups of patients with aphasia and with nonaphasic brain damage are not easy to evaluate.

De Renzi et al. (1966) have, in fact, reported highly significant defects on nonverbal abstraction tests in aphasic patients, but other authors (Meyers, 1948; Bauer & Beck, 1954; Zangwill, 1964) have failed to find a greater incidence of defects in aphasic patients than in comparable groups of nonaphasic patients with brain damage. The possibility that the difficulty of the task and other task characteristics may account for these discrepancies is supported by results recently obtained by Gainotti et al. (1986). These authors administered to normal controls, and to patients with aphasia and right-hemispheric damage two tests for "class inclusion" and "class intersection." No significant difference was observed between those with aphasia and those with right-hemispheric damage on the test of class inclusion, but aphasic patients performed significantly worse than any other control group the test of class intersection (see Table 5.1).

This discrepancy was tentatively explained by two different interpretations. The first assumed that the critical factor was the different difficulty of the two tasks: the former was relatively easy, the latter was rather difficult. However, the second interpretation assumed, that two different strategies could be used to solve the two apparently similar classification tasks, and that only one of these strategies really needed an "abstract attitude."

As a matter of fact, in the class inclusion test, the patients were presented with a certain number of colored pictures, all belonging to a given class (e.g., "castles") arranged in a large white circle representing the "class;" the patients were then given other pictures and instructed to select and put in the circle "all" and "only" those belonging to the class. Inasmuch as the members of the class unavoidably share common visual features, this test could be solved relying on purely perceptual criteria and did not necessarily require an abstract attitude.

On the class intersection test, however, a purely perceptual strategy was insufficient to solve the task, which consisted in presenting to the patients two sets of colored pictures, each representing a given class (e.g., "birds" = "class A" and "yellow objects" = "class B"). The pictures were arranged in linear order on two strips of paper, which were placed perpendicularly, with an intersection corresponding to one of their extremities; the patient was then requested to select (in a multiple-choice display) the figure belonging at the same time to both classes (e.g., "canary" = "bird" + "yellow"). This test could, therefore, be solved only by keeping simultaneously in mind the criteria of classification subsuming the two classes and searching for the picture fitting both the abstract criteria. Obviously these two interpretations would have different implications as for the "loss of the abstract attitude" hypothesis. If both tasks require the capacity to abstract the criterion that defines a given class, but only the most difficult one can significantly discriminate aphasic from nonaphasic patients, then the loss of the abstract attitude hypothesis is not supported. If, on the contrary, only the class intersection task selectively distinguishes aphasics from control patients, just because it cannot be solved without having resort to abstract operations, then the same hypothesis is upheld.

Unfortunately our data do not allow to choose between these alternative interpretations and we have reported them in some detail only to illustrate the difficulties of the problem and the need of exploring it with further, more carefully controled investigations.

The "Defective Analytical Isolation of Individual Features of Concepts" Hypothesis

The "defective analytical isolation of individual features of concepts hypothesis" recently proposed by Cohen and others (Cohen, Engel, Kelter, List, & Strohner, 1976; Kelter, Cohen, Engel, List, & Strohner, 1976; Cohen, Kelter, & Woll, 1980; Cohen & Woll, 1981; Cohen, Glockner, Lutz, Maier, & Meier, 1983) can be considered as a modern and more precise reformulation of the Goldstein view. Cohen's theory has two main advantages with respect to Goldstein's hypothesis: (*a*) it replaces the construct of "abstract attitude," which was probably too general and difficult to define operationally, with the more specific and operationally well-defined notion of "analytical processing of individual features of concepts;" and (*b*) it is well in line with recent views about the fundamental abilities underpinning the left hemisphere dominance for language, which was assumed by Bradshaw and Nettleton (1981) to imply a serial and analytical processing ability.

Cohen and associates advanced the hypothesis of a deficiency in the analytical isolation of individual features of concepts relying on the contrasting results obtained on two nonverbal matching tasks (Kelter et al., 1976; Cohen et al., 1980). Subjects had to decide in two different conditions which of two pictures was more closely related to a third target picture. In one condition the decision had to be based on the existence of a common situational context (e.g., guitar/violin—bullfight; Spain being the context mediator). In the other condition, the decision had to be based on the isolation and appreciation of critical features of the depicted objects (e.g., swan/turkey—snowman; white being the mediating individual feature).

Aphasic patients were as good as control subjects on the first type of task, but scored significantly worse than any other control group on the second task, which required the comparison of different concepts with respect to individual features. The researchers assumed that this basic deficiency in the appreciation of individual features of concepts could account not only for results obtained on nonverbal cognitive tasks, but also for results obtained on verbal tasks. They proposed, for example, that the high discriminating power between aphasic and nonaphasic patients shown by the Token test could be due to the analytical handling and isolation of features, such as color, size, or shape required by the test (Cohen et al., 1976). The notion of a selective deficiency in the analytical isolation and cognitive handling of individual features of concepts could also account for other kinds of nonverbal cognitive disturbances often observed in aphasic patients: the capacity to match typical colors with the corresponding drawings of objects, to match nonverbal meaningful sounds with the corresponding pictures, to mime the use of a presented object.

Unfortunately, results of a series of experimental investigations recently reported by Cohen et al. (1983) have failed to substantiate the hypothesis. Even in this case, further investigations are, therefore, needed to better control and define the notion of a selective impairment of analytical processing capacity in aphasia.

The "Conceptual Semantic Representation Deficit" Hypothesis

The conceptual/semantic representation deficit hypothesis can be considered as a reformulation (in terms of semantic representation) of Bay's suggestion (1962, 1964, 1974) that aphasia may be defined as a disorder in the organization of concepts and in the actualization of concepts in language. Our discussion of this theory will be necessarily very sketchy because a deeper analysis of this problem would require a better understanding of how concepts are represented in mind or, more generally, of the inner representations of the outside world. This is not possible because it is not clear if there is one central cognitive system in which our knowledge of the outside world is stored and organized, or if there are several modality-specific semantic systems. Furthermore, even the supporters of the latter hypothesis cannot specify if these independent semantic systems are or not hierarchically organized and how they reciprocally interact. In the present discussion we will, then, simply assume that concepts are formed by organized bundles of semantic (and perhaps also of perceptual) features, without specifying whether these features are structurally, or only functionally, interconnected.

The existence of a defect of the conceptual/semantic representation is strongly suggested by the observation that aphasic patients are severely impaired on nonverbal associative–conceptual tasks and by a qualitative analysis of the semantic–lexical aspects of their properly verbal aphasic disintegration. We have already seen that nonverbal cognitive disorders of aphasic patients basically consist in the inability to associate different features belonging to the concept of an object (such as its shape with its typical color, or the sound it usually produces, or the actions associated with it). If we consider that the organized association of these features allows the formation of concepts that are clearly demarcated and easily distinguishable from similar ones, we can easily admit that any loosening of associations (as revealed by nonverbal cognitive tasks) will give rise to frequent confusions between concepts having some main features in common. This sort of confusion is just what emerges from a qualitative analysis of the lexical–semantic errors of aphasic patients.

A careful analysis of these errors (cf. Rinnert and Whitaker, 1973; Buckingham & Rekart, 1979) has shown that a certain general knowledge of a word's meaning (such as the class to which the referent belongs) is usually retained in aphasic patients, but that some specific features are wrongly selected. The ensuing semantic confusion can reveal itself at the expressive level through the production of semantically related paraphasias, and at the receptive level through the inability to select the correct item in an array of semantically similar alternatives.

In addition to the qualitative aspects of the nonverbal cognitive impairment and of the semantic–lexical errors, the conceptual/semantic representation deficit hypothesis also accounts for the particularly strong association between these two orders of disturbances. It is, in fact, obvious that if a defect of the conceptual representation subsumes both nonverbal cognitive disturbances and verbal semantic–lexical defects, these two types of disorders should be usually observed in the same patients. Thus a lot of empirical data, collected without using an explicit model of conceptual representation, may be explained by the conceptual deficit hypothesis.

Some evidence in favor of this theory has also been recently obtained by Caramazza et al. (Whitehouse, Caramazza, & Zurif, 1978; Caramazza, Berndt, & Brownell, 1982), who used a more explicit model of conceptual representation developed by Labov (1973) and by Miller and Johnson-Laird (1976). In a first study (Whitehouse et al., 1978) two groups of Broca's and anomic patients were given a series of line drawings of food containers, varying on two dimensions: a continuum dimension of diameter-to-height ratio and a discrete dimension of presence/absence of a handle. Drawings were presented in different contexts (e.g., a coffee pot, a cereal box, a pitcher of ice water) and patients were requested to select the most appropriate name for the object from a multiple-choice list, including "cup," "bowl," and "glass." Normal controls and patients with Broca's aphasia were able to integrate in a predictable way perceptual and functional–contextual information in the selection of the most appropriate name for each stimulus; anomic patients were not reliably influenced by either perceptual or functional information in their choice of the verbal label. According to Whitehouse et al. (1978) the inability of anomic patients to classify the stimuli correctly suggests that naming difficulties reflect an impairment of the conceptual organization underlying word meaning.

Caramazza et al. (1982) tried to control this hypothesis by using the same stimulus material in a more strictly nonverbal task. They assumed that the perceptual parsing of an object (for classification, and eventually for naming) includes, after a preliminary low-level analysis, an active top–down component. In this second stage of the parsing, semantic considerations (namely the set of features that are represented as part of the concept of the object) may guide the perceptual analysis of the presented object. If the semantic component that guides the perceptual parsing is disrupted, patients should also be impaired in judging the perceptual similarity between pairs of visual stimuli, such as those used by Whitehouse et al. (1978).

The hypothesis was confirmed because aphasics patients who had failed to classify the stimuli under the correct verbal label were also impaired in the ability to perform perceptual similarity judgments for pairs of stimulus objects. Caramazza et al. (1982) therefore concluded that the semantic deficit hypothesis accounts for the co-occurrence of naming defects, nonverbal disorders of perceptual analysis, and object classification errors in patients with lesions to the posterior regions of the language areas.

There is, however, an important objection that can be addressed to a strong version of the conceptual deficit hypothesis. Aphasic patients are not only impaired on associative/conceptual tasks using stimuli that should logically have a well-defined conceptual representation (i.e., objects defined in the dictionaries with a list of perceptual, functional, or other abstract properties). They are also severely impaired in tasks of abstract thinking, such as the Weigl sorting test (Goldstein, 1948; De Renzi et al., 1966), which use, as material pieces of wood that do not have in their conceptual representation the criteria (e.g., color, form, size, thickness) used to classify them. The aphasic impairment on this task can hardly be ascribed to a loss of stored information and seems rather to point to the inability to actively search for the critical properties of objects, owing perhaps to disruption of the linguistic–semantic categories that assist the perceptual analysis.

Conclusions

Data discussed in the last section show that the relationships between verbal and nonverbal cognitive disorders of patients with aphasia cannot be exhaustively accounted for by any of the various interpretations advanced to explain this problem. Currently, the only position that is hardly disputable is that a functional relationship exists between verbal and nonverbal cognitive disturbances of aphasic patients. We are also in doubt that the verbal mediation hypothesis may be sufficiently powerful to explain the qualitative aspects of these cognitive disorders, inasmuch as this theory cannot help to clarify the nature of the semantic–lexical disorder intimately linked with the nonverbal cognitive disturbance. Thus, we are almost forced to admit that one of the main components of the aphasic symptom-complex is the disruption of a basic capacity, underlying both the lexical semantic and the nonverbal cognitive aspects of the aphasic syndromes. Many facts surveyed in the present review suggest that this basic defect could consist of either a disruption of the information stored in the conceptual representation or of an inability to fully access this reppresentation.

We would, however, conclude by saying that the preverbal basic defect hypothesis and the verbal mediation hypothesis are not necessarily opposed and incompatible. Indeed, it is possible that a basic preverbal disruption of the conceptual representation may directly account for some nonverbal associative disturbances and for the semantic–lexical disorders of aphasic patients. Naming disturbances, however, might in turn render even more difficult the execution of other nonverbal cognitive tasks, either disturbing the semantically–linguistically guided perceptual parsing of nonverbal stimuli, as proposed by Caramazza et al. (1982), or acting as thought distractors, as typically shown by patients with modality-specific aphasia. This second mechanism could, perhaps, help to explain the inability of aphasic patients to perform intelligence tests based on nonverbal abstract stimuli, which can hardly be accounted for by the conceptual representation deficit hypothesis.

Acknowledgment. The author is extremely grateful to Professor E. De Renzi for his generous help, advice and assistance in the preparation of this paper, which was supported by a Consiglio Nazionale delle Ricerche (National Research Council) grant.

References

Alajouanine, T. (1948). Aphasia and artistic realisation. *Brain*, *71*, 229–241.
Archibald, Y., Wepman, J., & Jones, L.V. (1967). Nonverbal cognitive perfor-

mance in aphasic and nonaphasic brain-damaged patients. *Cortex*, *3*, 275–294.
Arrigoni, G., & De Renzi, E. (1964). Constructional apraxia and hemispheric locus of lesion. *Cortex*, *1*, 170–197.
Assal, G., & Buttet, J. (1976). Couleurs, attributs d'objets. Leurs déficits lors des lésions du cortex cérébral. *Lyon Médical*, *236*, 685–689.
Basso, A., & Capitani, E. (1985). Spared musical abilities in a conductor with global aphasia and ideomotor apraxia. *Journal of Neurology, Neurosurgery, and Psychiatry*, *48*, 407–412.
Basso, A., De Renzi, E., Faglioni, P., Scotti, G., & Spinnler, H. (1973). Neuropsychological evidence for the existence of cerebral areas critical to the performance of intelligence tasks. *Brain*, *96*, 715–728.
Basso, A., Faglioni, P., & Spinnler, H. (1976). Non-verbal colour impairment of aphasics. *Neuropsychologia*, *14*, 183–193.
Bauer, K., & Beck, D. (1954). Intellect after cerebro-vascular accident. *Journal of Nervous and Mental Disease*, *120*, 379–395.
Bay, E. (1962). Aphasia and non-verbal disorders of language. *Brain*, *85*, 411–426.
Bay, E. (1964). Principles of classification and their influence on our concepts of aphasia: In A.V.S. De Reuck & M. O'Connor (Eds.), *Disorders of language* (pp. 122–142). London: Ciba Foundation Symposium–Churchill.
Bay, E. (1974). Intelligence and aphasia: In Y. Lebrun & R. Hopps (Ed.), *Intelligence and Aphasia*. Amsterdam: Swets & Zeitlinger.
Beauvois, M.F. (1982), Optic aphasia: a process of interaction between vision and language. *Philosophical Transactions of the Royal Society. Lodon B298*, 35–47.
Beauvois, M.F., Saillant, B., Meininger, V., & Lhermitte, F. (1978). Bilateral tactile aphasia: a tacto-verbal dysfunction. *Brain*, *101*, 381–401.
Benton, A.L. (1973). Visuoconstructive disability in patients with cerebral disease: Its relationship to side of lesion and aphasic disorder. *Documenta Ophthalmologica*, *34*, 67–76.
Bradshaw, J.L., & Nettleton, N.C. (1981). The nature of hemispheric specialization in man. *Behavioral and Brain Sciences 4*, 51–91.
Broca, P. (1861). *Bulletin de la Société d'Anatomie* Remarques sur le siège de la faculté du language articulé, suivies d'une observation d'aphémie (perte de la parole). Bull, Sté Anat. Paris, *6*, 330–357.
Buckingham, H.V., & Rekart, D.M. (1979). Semantic paraphasia. *Journal of Communication Disorders*, *12*, 197–209.
Caramazza, A., Berndt, R.S., & Brownell, H.H. (1982), The semantic deficit hypothesis: perceptual parsing and object classification by aphasic patients. *Brain and Language*, *15*, 161–189.
Christopoulou, C., & Bonvillian, J.D. (1985). Sign language, pantomime, and gestural processing in aphasic persons: A review. *Journal of Communications Disorders*, *18*, 1–20.
Cicone, M., Wapner, W., Foldi, N., Zurif, E., & Gardner, H. (1979), The relation between gesture and language in aphasic communication. *Brain and Language*, *8*, 324–349.
Cohen, R., Glöckner, A., Lutz, M., Maier, T., & E. Meier (1983). Cognitive impairments in aphasia: new results and new problems. In R. Bäuerle, C. Schwarze, & A. von Stechow (Eds.), *Meaning, Use, and Interpretation of Language* (pp. 30–45). Berlin: Walter de Gruyter.
Cohen, R., & Kelter, S. (1979). Cognitive impairment of aphasics in a colour-to-picture matching task. *Cortex*, *15*, 235–245.

Cohen, R., Engel, D., Kelter, S., List, G., & Strohner, H. (1976). Restricted associations in aphasics and schizophrenics. *Archiv für Psychiatrie und Nervenkrankheiten*, *222*, 325–338.

Cohen, R., Kelter, S., & Woll, G. (1980), Analytical competence and language impairment in aphasia. *Brain and Language*, *10*, 331–347.

Cohen, R., & Woll, G. (1981), Facets of analytical processing in aphasia: A picture ordering task. *Cortex*, *17*, 557–570.

Colonna, A., & Faglioni, P. (1966). The performance of hemisphere damaged patients on spatial intelligence tests. *Cortex*, *2*, 293–307.

Daniloff, J.K., Noll, J.D., Fristoe, M., & Lloyd, L.L. (1982). Gesture recognition in patients with aphasia. *Journal of Speech and Hearing Disorders*, *47*, 43–49.

De Renzi, E., & Faglioni, P. (1965). The comparative efficiency of intelligence and vigilance tests in detecting hemispheric cerebral damage. *Cortex*, *1*, 410–433.

De Renzi, E., Faglioni, P., Savoiardo, M., & Vignolo, L.A. (1966). The influence of aphasia and of the hemispheric side of the cerebral lesion on abstract thinking. *Cortex*, *2*, 399–420.

De Renzi, E., Faglioni, P., Scotti, G., & Spinnler, H. (1972). Impairment in associating colour to form, concomitant with aphasia. *Brain*, *95*, 293–304.

De Renzi, E., Pieczuro, A., & Vignolo, L.A. (1968). Ideational apraxia: A quantitative study. *Neuropsychologia*, *6*, 41–52.

De Renzi, E., Scotti, G., & Spinnler, H. (1969). Perceptual and associative disorders of visual recognition: relationship to the side of the cerebral lesion. *Neurology*, *19*, 634–642.

De Renzi, E., & Spinnler, H. (1967). Impaired performance on color tasks in patients with hemispheric damage. *Cortex*, *3*, 194–217.

Doehring, D.G., Dudley, J.G., & Coderre, L. (1967). Programmed instruction in picture-sound association for the aphasic. *Folia Phoniatrica*, *19*, 414–426.

Duffy, R.J. (1984). Comparison of the performance of a fluent and a nonfluent aphasic on a pantomimic referential task. *Brain and Language*, *21*, 260–273.

Duffy, R.J., & Duffy, J.R. (1981). Three studies of deficits in pantomime expression and pantomimic recognition in aphasia. *Journal of Speech and Hearing Research*, *24*, 97–111.

Duffy, R.J., Duffy, J.R., & Mercaitis, P.A. (1984). Comparison of the performances of a fluent and a nonfluent aphasic on a pantomimic referential task. *Brain and Language*, *21*, 260–273.

Duffy, R.J., Duffy, J.R., & Pearson, R.L. (1975). Impairment of pantomime recognition in aphasics. *Journal of Speech and Hearing Research*, *18*, 115–132.

Duffy, R.J., & Liles, B.Z. (1979). A translation of Finkelnburg's (1870) lecture on aphasia as 'asymboly' with commentary. *Journal of Speech and Hearing Disorders*, *44*, 156–168.

Duffy, R.J., & McEwen, W.J. (1978). A study of the relationship between pantomime symbolism and pantomime recognition in aphasics. *Folia Phoniatrica*, *30*, 286–292.

Duffy, J.R., & Watkins, L.B. (1984). The effect of response choice relatedness on pantomime and verbal recognition ability in aphasic patients. *Brain and Language*, *21*, 291–306.

Elithorn, A. (1955). A preliminary report on a preceptual maze test sensitive to brain damage. *Journal of Neurology, Neurosurgery and Psychiatry*, *18*, 287–292.

Faglioni, P., Spinnler, H., & Vignolo, L.A. (1969). Contrasting behavior of right and left hemisphere-damaged patients on a discriminative and a semantic task of

auditory recognition. *Cortex*, *5*, 366–389.
Finkelnburg, D.C. (1870). Niederrheinische Gesellschaft, Sitzung vom 21. Marz 1870 in Bonn. *Berliner Klinische Wochenschrift*, *7*, 449–450, 460–462.
Gainotti, G., Caltagirone, C., & Ibba, A. (1975). Semantic and phonemic aspects of auditory language comprehension in aphasia. *Linguistics*, *154/155*, 15–29.
Gainotti, G., Carlomagno, S., Craca, A., & Silveri, M.C. (1986). Disorders of classificatory activities in aphasia. *Brain and Language*, 28, 181–195.
Gainotti, G., D'Erme, P., Villa, G., & Caltagirone, C. (1986). Focal brain lesions and intelligence: a study with a new version of Raven's colored matrices. *Journal of Clinical and Experimental Neuropsychology*, *8*, 37–50.
Gainotti, G., & Ibba, R. (1972). La comprehensione del significato dei gesti simbolici negli afasici. *Minerva Psichiatrica e Psicologica*, *13*, 1–9.
Gainotti, G., & Lemmo, M.A. (1976). Comprehension of symbolic gestures in aphasia. *Brain and Language*, *3*, 451–460.
Gainotti, G., Miceli, G., Caltagirone, C. (1979). The relationships between conceptual and semantic-lexical disorders in aphasia. *International Journal of Neuroscience*, *10*, 45–50.
Gainotti, G., Nocentini, U., Sena, E., & Silveri, M.C. (1986). Discovery of simple binary sequences in brain-damaged patients. *International Journal of Clinical Neuropsychology*, *8*, 99–103.
Gainotti, G., Silveri, M.C., Villa, G., & Caltagirone, C. (1983). Drawing objects from memory in aphasia. *Brain*, *106*, 613–622.
Gelb, A., & Goldstein, K. (1924). *Psychologische Analysen Hirnpathologischer Falle*. Leipzig: Barth.
Geschwind, N., & Fusillo, M. (1966). Color naming defects in association with alexia. *Archives of Neurology*, *15*, 137–146.
Gil, R., Pluchon, C., Toullat, G., Micheneau, D., Rogez, R., & Lefevre, J.P. (1985). Disconnexion visuo-verbale (aphasie optique) pour les objets les images, les couleurs et les visages avec alexie "abstractive." *Neuropsychologia*, *23*(3), 333–349.
Glass, A.V., Gazzaniga, M.S., & Premack, D. (1973). Artificial language training in global aphasics. *Neuropsychologia*, *11*, 95–103.
Goldstein, K. (1948). *Language and Language Disturbances*. New York: Grune & Stratton, pp. 246–291.
Goldstein, K., & Scheerer, M. (1941). Abstract and concrete behavior: An experimental study with special tests. *Psychological Monographs*, *43*, 1–151.
Goodglass, H., & Kaplan, E. (1963). Disturbance of gesture and pantomime in aphasia. *Brain*, *86*, 703–720.
Head, H. (1926). *Aphasia and kindred disorders of speech* (vol. 1). Cambridge, MA: Cambridge University Press.
Jackson, J.H. (1878). On affections of speech from disease of the brain. *Brain*, *1*, 304–330.
Kelter, S., Cohen, R., Engel, D., List, G., & Strohner, H. (1976). Aphasic disorders in matching tasks involving conceptual analysis and covert naming. *Cortex*, *12*, 383–394.
Kussmaul, A. (1877). Disturbances of speech: an attempt in the pathology of speech: In H.v. Ziemssen (Ed.), *Cyclopedia of the Practice of Medicine, Vol. 14*. New York: Wood.
Laborde (1863). Discussion sur la communication de Parrot. Bulletin de la Société

de Anatomie. 8, 423. Quoted by: Ombredane, A. (1951) *L'aphasie et l'élaboration de la pensée explicite*, P.U.F. (Paris).

Labov, W. (1973). The boundaries of words and their meanings. In C. Bailey and R. Shuy (Eds). *New Ways of Analyzing Variation in English*. Washington D.C.: Georgetown University Press.

Lhermitte, F., & Beauvois, M.F. (1973). A visual–speech disconnection syndrome. Report of a case with optic aphasia, agnosic alexia and colour agnosia. *Brain*, *96*, 695–714.

Liepmann, H. (1905). Der weitere Krankheitsverlauf bei dem einseiting Apraktischen und der Gehirnbefund auf Grund von Serienschnitten. *Monatschrift für Psychiatrie und Neurologie*, *17*, 289.

Marie, P. (1926). La troisième circonvolution frontale gauche ne joue aucun role special dans la fonction du language. *Semaine Médicale*, 1906, May 23. Reprinted in Trauvaux et memoires. Vol. I Paris: Masson.

Meyers, R. (1948). Relation of "thinking" and language: an experimental approach using dysphasic patients. *Archives of Neurology and Psychiatry* (Chicago), *60*, 119–139.

Miller, G.A., & Johnson-Laird, P.N. (1976). *Language and Perception*. Cambridge: Cambridge University Press, pp. 226–230.

Netsu, R., & Marquardt, T.P. (1984). Pantomime in aphasia: Effects of stimulus characteristics. *Journal of Communication Disorders*, *17*, 37–46.

Peterson, L.N., & Kirshner, H.S. (1981). Gestural impairment and gestural ability in aphasia: A review. *Brain and Language*, *14*, 333–348.

Pickett, L.W. (1974). An assessment of gestural and pantomimic deficit in aphasic patients. *Acta Symbolica*, *5*, 69–86.

Piras, M.R., De Vreese, L., Ortu, R., D'Onofrio, M., & Agnetti, V. (1985). Afasia ottica, alessia pura senza agrafia, anomia per i colori: un processo di interazione fra visione e linguaggio. Communication held at the 24th Congress of the Italian Society of Neurology (Pisa, October 10, 1985).

Premack, D. (1971). Language in chimpanzee? *Science*, *172*, 808–822.

Raven, J.C. (1938). *Standard Progressive Matrices*, London: H.K. Lewis.

Raven, J.C. (1962). *Coloured Progressive Matrices Sets A, Ab, B*. London: H.K. Lewis (originally published in 1947, revised order in 1956).

Raven, J.C. (1965). *Guide to Using the Coloured Progressive Matrices Sets A, Ab, B*. London: H.K. Lewis.

Rinnert, C., & Whitaker, H.A. (1973). Semantic confusions by aphasic patients. *Cortex*, *9*, 56–81.

Russo, M., & Vignolo, L.A. (1967). Visual figure–ground discrimination in patients with unilateral cerebral disease. *Cortex*, *3*, 113–127.

Semmes, J., Weinstein, S., Ghent, L., & Teuber, H.-L. (1954). Performance on complex tactual tasks after brain injury in man: Analysis by locus of lesion. *American Journal of Psychology*, *67*, 220.

Seron, X., Van Der Kaa, M.A., Remitz, A., & Van Der Linden, M. (1979). Pantomime interpretation and aphasia. *Neuropsychologia*, *17*, 661–668.

Spinnler, H., & Vignolo, L.A. (1966). Impaired recognition of meaningful sounds in aphasia. *Cortex*, *2*, 337–348.

Strohner; H., Cohen, R., Kelter, S., & Woll, G. (1978). "Semantic" and "acoustic" errors of aphasic and schizophrenic patients in a sound-picture matching task. *Cortex*, *14*, 391–403.

Teuber, H.L., & Weinstein, S. (1956). Ability to discover hidden figures after cerebral lesions. *Archives of Neurology and Psychiatry*, *76*, 369–379.
Trousseau, A. (1865). Clinique Médicale de l'Hotel Dieu de Paris, T. II, Paris.
Tulving, E. (1972). Episodic and semantic memory. In E. Tulving & W. Donaldson (Eds.), *Organization and Memory* (pp. 381–403). New York, Academic Press.
Van Dongen, H.R. (1973). Impairment of drawing and intelligence in aphasic patients. Unpublished paper presented at the Brussels Conference on Intelligence and Aphasia. Summarized in Y. Lebrun & R. Hoops (Eds), *Intelligence and Aphasia*. Amsterdam: Swets & Zeitlinger.
Varney, N.R. (1978). Linguistic correlates of pantomime recognition in aphasic patients. *Journal of Neurology, Neurosurgery and Psychiatry*, *41*, 564–568.
Varney, N.R. (1980). Sound recognition in relation to aural language comprehension in aphasic patients. *Journal of Neurology, Neurosurgery and Psychiatry*, *43*, 71–75.
Varney, N.R. (1982). Pantomime recognition defect in aphasia: Implications for the concept of asymbolia. *Brain and Language*, *15*, 32–39.
Varney, N.R., & Benton, A.L. (1982). Qualitative aspects of pantomime recognition deficit in aphasia. *Brain and Cognition*, *1*, 132–139.
Varney, N.R., & Vilensky, J.A. (1980). Neuropsychological implications for preadaptation and language evolution. *Journal of Human Evolution, 9*, 223–226.
Vygotsky, L.S. (1962). *Thought and language*. E. Haufmann & G. Vaker (Eds.), Cambridge, MA: MIT Press.
Wapner, W., & Gardner, H. (1981). Profiles of symbol-reading skills in organic patients. *Brain and Language*, *12*, 303–312.
Weigl, E. (1927). Zur Psychologie sogenannter Abstraktionsprozesse. *Zeitschrift für Psychologie*, *103*, 2–45.
Weinstein, S. (1964). Deficits concomitant with aphasia and lesions of either cerebral hemisphere. *Cortex*, *1*, 154–169.
Weinstein, S., Teuber, H.L., Ghent, L., & Semmers, J. (1955). Complex visual performance after penetrating brain injury in man. *American Psychologist*, *10*, 408.
Weisenburg, T.H., & McBride, K.E. (1935). *Aphasia: a clinical and psychological study*. New York, Commonwealth Fund Hildred and Co.
Whitehouse, P., Caramazza, A., & Zurif, E.B. (1978). Naming in aphasia: Interacting effects of form and function. *Brain and Language*, *6*, 63–74.
Zangwill, O.L. (1964). Intelligence in aphasia. In A.V.S. De Reuck & M. O'Connor (Eds.), *Disorders of Language* (pp. 261–274). Ciba Foundation Symposium. London: Churchill.
Zangwill, O.L. (1969). Intellectual status in aphasia. In P.J. Vinken & G.W. Bruyn (Eds.) *Handbook of Clinical Neurology* (Vol. 4). (pp. 105–111). Amsterdam: North-Holland Publishing.

Author Index

Subject Index